Collins

CSEC® GEOGRAPHY ATLAS SKILLS WORKBOOK

Naam Thomas
Reviewers: Sheridon King Coke and Monique Campbell

Collins

William Collins' dream of knowledge for all began with the publication of his first book in 1819. A self-educated mill worker, he not only enriched millions of lives, but also founded a flourishing publishing house. Today, staying true to this spirit, Collins books are packed with inspiration, innovation and practical expertise. They place you at the centre of a world of possibility and give you exactly what you need to explore it.

Collins. Freedom to teach.

Published by Collins
An imprint of HarperCollins*Publishers*
Westerhill Road
Bishopbriggs
Glasgow G64 2QT

HarperCollins*Publishers*
Macken House, 39/40 Mayor Street Upper, Dublin 1, D01 C9W8, Ireland

Browse the complete Collins Caribbean catalogue at
www.collins.co.uk/caribbeanschools

© HarperCollins*Publishers* Limited 2021

10 9 8 7 6 5 4

ISBN 978-0-00-842013-0

Collins **CSEC®** Geography Atlas Skills Workbook is an independent publication and has not been authorised, sponsored or otherwise approved by **CXC®**.

CSEC® is a registered trademark of the **Caribbean Examinations Council (CXC®)**.

All rights reserved. No part of this publication may be reproduced, stored in a retrieval system, or transmitted in any form by any means, electronic, mechanical, photocopying, recording or otherwise, without the prior written permission of the Publisher or a licence permitting restricted copying in the United Kingdom issued by The Copyright Licensing Agency Ltd, 5th Floor, Shackleton House, 4 Battle Bridge Lane, London, SE1 2HX.

British Library Cataloguing in Publication Data
A catalogue record for this publication is available from the British Library.

The publishers gratefully acknowledge the permission granted to reproduce the copyright material in this book. Every effort has been made to trace copyright holders and to obtain their permission for the use of copyright material. The publishers will gladly receive any information enabling them to rectify any error or omission at the first opportunity.

Author: Naam Thomas
Author of the SBA Section: Sheridon King Coke
Reviewers: Sheridon King Coke and Monique Campbell
Publisher: Dr Elaine Higgleton
In-house senior editor: Julianna Dunn
Project manager: Sonia Dawkins
Copy editor: Mitch Fitton
Proofreader: Helen Bleck
Typesetter: QBS Learning
Mapping and cover design: Gordon MacGilp
Production controller: Lyndsey Rogers
Printed and bound in the U.K.

This book contains FSC™ certified paper and other controlled sources to ensure responsible forest management.

For more information visit: www.harpercollins.co.uk/green

Contents

Introduction 4

SECTION 1: MAPWORK AND MAP SKILLS
Map symbols and types 5
Compass directions and bearings 8
Latitude and longitude............................. 10
Time zones.. 15
Grid references 17
Sketch maps... 20
Scale 22
Cross-sections 24
Large scale mapwork (Ordnance survey maps):
   St Vincent: La Soufrière volcano 26
Barbados: Cove Bay................................ 28
Trinidad: Port of Spain............................ 30
Trinidad: San Fernando.......................... 32

SECTION 2: NATURAL SYSTEMS
Internal structure of the Earth............... 34
Earthquakes: processes.......................... 40
Volcanoes: structure and formation..... 42
Volcanoes: volcanic landscapes............ 48
Weathering.. 50
Mass movement 54
Limestone: characteristics and processes 57
Weather and climate 61
Global warming and climate change.... 70
Ecosystems 77
Soil... 86
The water cycle and fluvial systems.... 90
Coastal systems 99
Natural hazards 111

SECTION 3: HUMAN SYSTEMS
Population distribution 117
Caribbean population distribution.... 121
Population structure 125
Population change 128
Population growth: Caribbean 133
Population: China................................. 135
Population: Nigeria............................... 137
Urbanisation 139
Urbanisation: Kingston........................ 141
Migration.. 143
Migration: Jamaica............................... 146
Caribbean resources 148
Economic sectors 150
Jamaican resources: bauxite............. 154
Trinidad resources: oil and gas......... 157
Primary industry: agriculture 160
Primary industry: fishing in Belize 162
Primary industry: forestry in Guyana 165
Secondary industry: food processing in
   Singapore and CARICOM 168
Tertiary industry: tourism................... 171

The School Based Assessment.......... 175

Introduction

This workbook has been written to meet the need for practice materials that help students develop the skills they require for **CSEC®** Geography. Teachers will also find the activities useful for assessing how well students are doing and for identifying areas where students need more support. Many of the activities in this workbook are based on a review of the examiners' reports and an understanding of the problems that students have with geography at **CSEC®**. This workbook should be used in association with the Collins Student Atlas for the Caribbean and can be used to support any scheme of study and to complement any textbook or study guide. Using the atlas as the backbone for exploring the syllabus provides students with an abundance of familiar visual cues that make for a smooth transition from lower secondary into **CSEC®** study.

For the student, exercises are designed to strengthen foundation skills introduced in early secondary school while challenging them to develop new skills appropriate for the **CSEC®** level. A key feature aimed at successful exam-taking is a writing development strategy that asks students to build their responses around listed keywords. Students can practise the use of these key terms – which they must use in an appropriate and accurate way to score good marks in the exam – through a range of writing activities. Diagrams, drawings and written work are presented side-by-side with practical projects and real-world applications. This removes the isolation of rote learning and instead promotes associative learning. This associative learning approach – coupled with the wide variety of activities and exercises – ensures that key objectives from the syllabus can be taught, studied and assessed in a classroom of mixed ability and with a variety of learning styles.

For the teacher, this workbook provides methods for assessing elementary and higher-order thinking skills including:

- Descriptive and observational questions which are linked to colourful and detailed graphs, charts and maps in the Collins Student Atlas for the Caribbean.

- Immediate recall and labelling exercises with culturally relevant and well-known contemporary examples which keep students engaged and can be used for a change of pace during class.

- Natural systems exercises which require that students draw a series of diagrams to explain processes, focusing on visualisation.

- Sociological and historical case studies, encouraging students to practise extracting relevant facts from reports to build a useful picture of topics in human systems.

- Cross referencing exercises which ask the student to reference multiple maps, graphics or data to draw a well-supported conclusion on a single topic. This zooms in on analytical ability.

- Research questions aimed at building students' ability to discriminate and identify high quality sources – these are a particularly useful feature of the workbook.

- Special attention is paid to using Caribbean examples and locations so that students are inspired to think creatively about finding Caribbean solutions to Caribbean problems.

- **CXC**-style questions are used extensively to acclimate students to the style.

The School Based Assessment guidance section of the workbook offers clear guidance to take students through each part of a research project in a step-by-step fashion that simplifies what can be an otherwise intimidating part of the requirements at this level. This streamlining, backed by sensible, practical suggestions for field work, means that this guidance creates a strong foundation for any research, the principles of which can also be used at higher levels of study.

Maps are used throughout the workbook as the Atlas is the key reference publication. A separate mapwork section is also included to allow students to practise the full scope of mapwork skills required for the **CSEC®** syllabus and exam. Special efforts were made to include ordnance survey maps and related questions as these are the map types used in the **CSEC®** Geography examination and are therefore an invaluable tool.

SECTION 1: MAPWORK AND MAP SKILLS
Map symbols and types

Study page 6 of your atlas.

1 Identify the map type shown on the following pages of your atlas.

ATLAS PAGE	MAP TYPE
41 – Jamaica: Population density, 2011	
57 – Barbados	
66 – Trinidad and Tobago: Population (top left)	
78 – North America: Climate July temperature	
108 – Africa: Relief	
141 – World: Political	

2 Study the *Dominica* map on page 52 of your atlas to complete the following exercise.

(a) In which parish is Boeri Lake located? _____

(b) In which parish is the main airport located? _____

(c) List the FOUR parishes where Morne Diablotins National Park is located.
_____

(d) Name ONE river on which a waterfall is located and the parish that the river is located in.
_____

3 **Study page 84 of the atlas to answer the following questions.**

Each map represents data about the population of the United States of America (USA), in a different way.

(a) The two top maps both represent population data.

   (i) What type of map is the *Population per sq km* map?

   _____

   (ii) What type of map is the *Urban agglomerations* map?

   _____

(b) (i) What is the lowest population number that can be represented on the *Urban agglomerations* map? _____

   (ii) State, considering your answer in **b(i)** above, ONE advantage of using the *Population per sq km* map instead of the *Urban agglomerations* map, to retrieve population data.

   _____

(c) (i) What is the highest population number that can be represented on the *Population per sq km* map? _____

   (ii) Explain, considering your answer in **c(i)** above, which map is more useful for showing very large population numbers.

   _____

   _____

4 Study the *Puerto Rico: Features* map on page 47 of your atlas to draw the correct symbol for each feature listed:

FEATURE	SYMBOL	FEATURE	SYMBOL
Point of interest		Main airport	
Port		Major marina	

6

5 Study the *Trinidad: Port of Spain* map key on page 68 of your atlas to draw the correct symbol for each feature listed. Use appropriate colours for each symbol.

FEATURE	SYMBOL	FEATURE	SYMBOL
Boundary - Forest		Reservoir	
Road - 2nd Class		Dam	
Trigonometrical Station (Minor)		Coconut	
Contours		Cemetery	
Hospital		Swamp	

6 Study the *Caribbean: Resources, Energy and Minerals* map on page 28 of your atlas to answer the following questions:

(a) List TWO countries where oilfields are located. _____

(b) List the countries where nickel is located. _____

(c) State how many oil refineries are shown on the map. _____

7

Compass directions and bearings

1) Study pages 18 and 19 of your atlas to complete the following table. State the correct 16-point compass directions of the locations in column B from those in column A.

A	B	COMPASS DIRECTION
Bonaire	Grenada	
Montserrat	Barbuda	
Great Abaco	New Providence	
Barbados	Tobago	
Martinique	St Croix	
Cerro de Punta (1338)	Milwaukee Deep (8605)	
Pic la Selle (2680)	Pico Turquino (1994)	
Mt Roraima (2810)	Cerro Curutú (1800)	

2) Study the *St Lucia: Features* map on page 53 of your atlas. Starting at the Edmund Forest Reserve, in which 16-point compass direction does each point of interest listed lie?

POINT OF INTEREST	COMPASS DIRECTION
Anse Chastanet Marine National Park	
La Fargue Craft Centre	
Latille Waterfalls and Gardens	
Savannes Bay Nature Reserve	
Frigate Islands Nature Reserve	
Fort Rodney	
Roseau Valley Banana Plantation	
Beausejour Cricket Stadium	

3 Study the *Grenada* map on page 56 of your atlas.

(a) Calculate the bearing of each peak listed below from Mount Qua Qua.

(i) Mount St Catherine _____

(ii) Mount Lebanon _____

(iii) Mount Sinai _____

(iv) Mount Granby _____

(b) Study the *Grenadines* map at the bottom right of page 54. Calculate the bearing of each town listed from Derrick, Bequia.

(i) Port Elizabeth, Bequia _____

(ii) Lovell Village, Mustique _____

(iii) Charlestown, Canouan _____

(iv) Ashton, Union Island _____

(c) Look at a map of your country. Calculate the bearing of each of the following locations from your school. Provide the names of the locations where appropriate.

LOCATION	BEARING
Capital city: _____	
The main airport: _____	
The closest secondary school: _____	
The major hospital: _____	
The main harbour (seaport): _____	
Your home: _____	

Latitude and longitude

1 Study page 8 of your atlas to complete the following exercise.

(a) Define latitude. _____

_____

(b) Define longitude. _____

_____

(c) State ONE similarity between the Equator and the Greenwich Meridian.

_____

_____

(d) State ONE difference between the lengths of the lines of longitude and the lines of latitude. _____

_____

(e) Describe the relationship between degrees and minutes in terms of the lines of latitude and longitude. _____

_____

_____

(f) Identify the error in each of the following sentences about the rules for using latitude and longitude. Rewrite each statement correctly.

  (i) Longitude is always stated before latitude. _____

  _____

  (ii) Latitude is read starting at the Greenwich Meridian. _____

  _____

  (iii) Degrees of longitude are numbered from 0° to 90°. _____

  _____

(g) Mark the following places on the diagram of the Earth; make sure you label the relevant degrees of latitude:

| Equator | Tropic of Cancer | Tropic of Capricorn | North Pole |
| South Pole | Arctic Circle | Greenwich Meridian | |

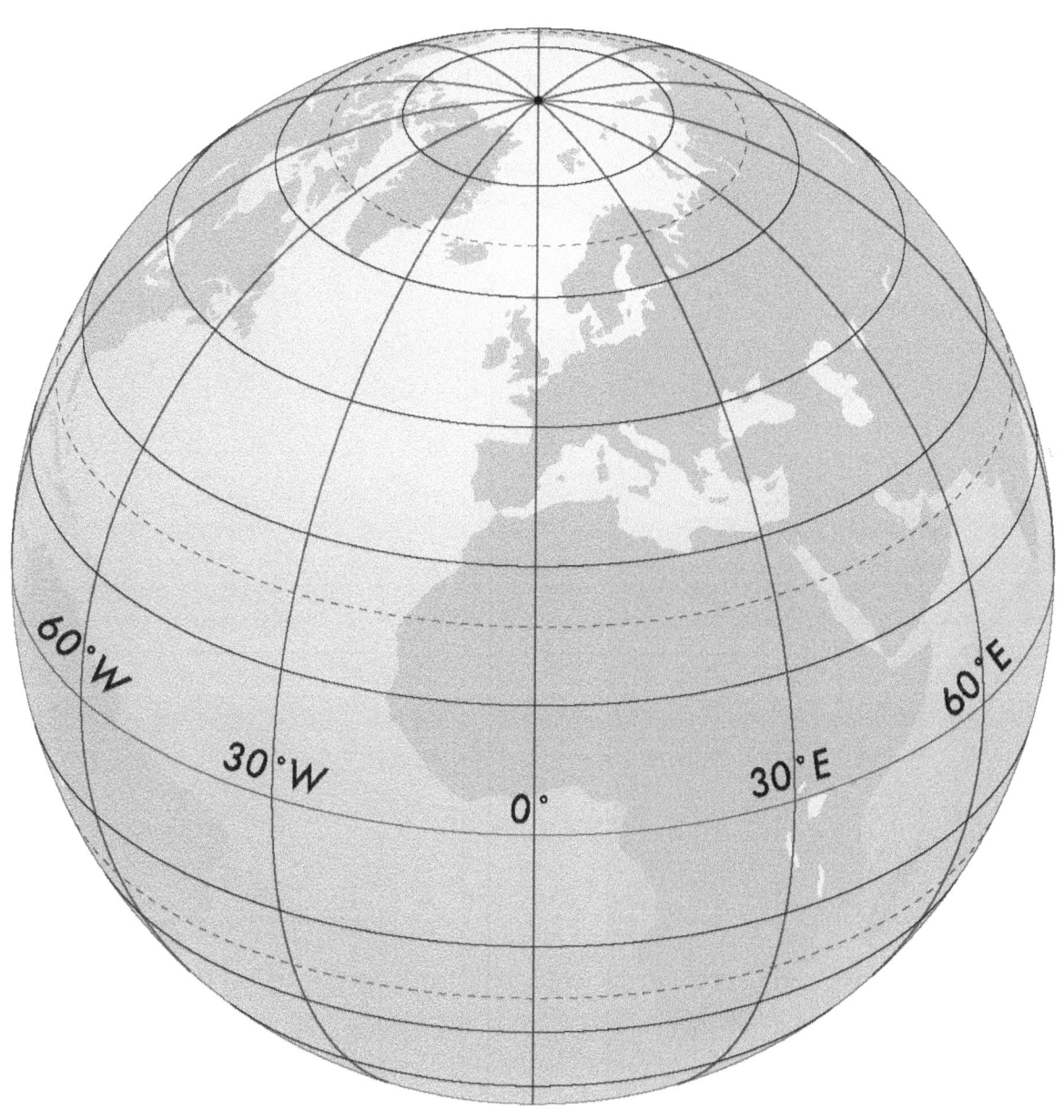

2 Study page 8 and the *World: Political* map on pages 140 and 141 of your atlas. Identify the Equator and the Greenwich Meridian. Organise the following countries into the correct areas on the diagram below to show their relative locations.

| Mongolia | Australia | Brazil | Nicaragua | Madagascar | Tunisia |
| Sri Lanka | Colombia | Cuba | Germany | Norway | Canada |

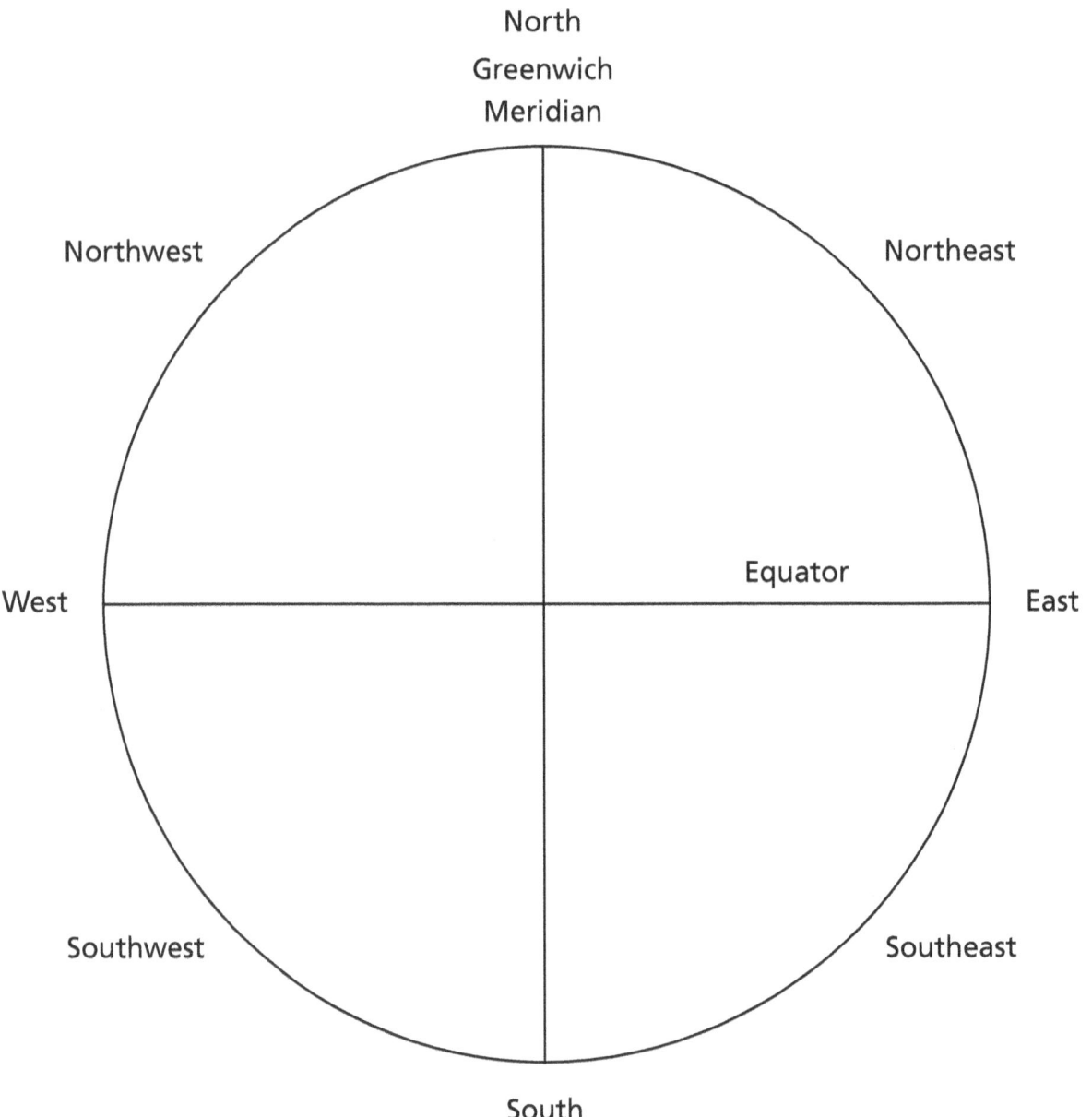

3 Study the political map of the Caribbean on pages 20 and 21 to determine the latitude and longitude for the capital cities of the following Caribbean countries.

(a) Nassau, The Bahamas _____

(b) Havana, Cuba _____

(c) Kingston, Jamaica _____

(d) Port-au-Prince, Haiti _____

(e) Basseterre, St Kitts and Nevis _____

(f) St John's, Antigua _____

(g) Castries, St Lucia _____

(h) St George's, Grenada _____

(i) Port of Spain, Trinidad and Tobago _____

(j) Georgetown, Guyana _____

4 Use the *Caribbean: Relief* map on pages 18 and 19 of your atlas, to locate and correctly name the mountain peaks found at each following latitude and longitude. Also, state the name of the country where each highland is found.

(a) 18°N 76°W _____

(b) 17°N 88°W _____

(c) 5°N 61°W _____

(d) 3°N 56°W _____

(e) 19°N 71°W _____

5 Use the *World: Political* map on pages 140 and 141 of your atlas to find and circle the error in each coordinate of latitude and longitude provided for the following international cities. Write down the correct location.

(a) Washington D.C., USA 38°N ~~87°W~~
    Correct location _____ 38°N 77°W

(b) Panama City, Panama 8°N ~~79°E~~
    Correct location _____ 8°N 79°W

(c) Montevideo, Uruguay 34°N 56°W
    Correct location _____

(d) Monrovia, Liberia ~~26°N~~ 10°W
    Correct location _____ 6°N 10°W

(e) Athens, Greece 37°N ~~19°E~~
    Correct location _____ 37°N 23°E

(f) Addis Ababa, Ethiopia ~~0°N~~ 38°E
    Correct location _____ 9°N 38°E

(g) Manila, Philippines ~~120°N~~ 120°E
    Correct location _____ 14°N 120°E

(h) Bangkok, Thailand 13°N ~~100°W~~
    Correct location _____ 13°N 100°E

Time zones

1 Study page 11 of your atlas on *Time Zones*. Use the time zone map to complete the table below. Use 24-hour time notations in your responses, for example 6:00 pm would be 18:00 in 24-hour time notation.

GREENWICH MEAN TIME (GMT)	CAPITAL CITY	CITY TIME (24-HOUR)
12:00	Anchorage, Alaska	
8:34	Denver, Colorado, USA	
4:30	Caracas, Venezuela	
13:22	Dakar, Senegal	
10:20	Oslo, Norway	
16:30	Cairo, Egypt	
19:10	Chengdu, China	
00:00	Sydney, Australia	

2 Complete the table below by calculating the Standard Time for the longitudes at 12:00 pm GMT. Use 12-hour time notations in your responses. Remember that 12-hour time notations must include 'am' and 'pm'.

LONGITUDE	STANDARD TIME (12-HOUR)
4° East	*12:16 pm*
115° East	
12.5° East	
97.5° East	
23° West	
177° West	
65° West	
50° West	

3 Study the *World: Relief* map on pages 142 and 143 of your atlas. Complete the table below by locating each mountain peak listed and noting its longitude, then calculate the Standard Time at the mountain peak at 12:00 pm GMT. Use 24-hour time notation in your responses as shown in the completed entry.

MOUNTAIN PEAK	LONGITUDE	STANDARD TIME (24-HOUR)
Mt Logan, Canada	*140° W*	*02:42*
Kilimanjaro, Tanzania		
Mont Blanc, France		
Chimborazo, Ecuador		
K2, Pakistan		
El'brus, Russia		
Mt Everest, China/Nepal		
Mt Whitney, USA		

Grid references

1 Study the *Guyana: Georgetown* map on page 72 of your atlas. State the four-figure grid reference for each of the following locations.

(a) Malgre Tout

_____

(b) Vauxhall

_____

(c) Sage Pond

_____

(d) Ruimveldt Industrial Estate

_____

(e) Air photo principal point 180/GY/1 040 (**HINT:** check the key).

_____

2 Refer to the *Trinidad: Port of Spain* map on page 68 of your atlas. Give the four-figure and six-figure grid references of each of the following locations.

(a) The Tourist Board building, on the eastern end of King's Wharf, Port of Spain

four-figure _____ six-figure _____

(b) The Cathedral opposite Woodford Square (Cathedral of Immaculate Conception)

four-figure _____ six-figure _____

(c) The Reservoir west of Fort Picton

four-figure _____ six-figure _____

(d) The Sewage Plant just off of Beetham Highway

four-figure _____ six-figure _____

(e) The mouth of Caroni River

four-figure _____ six-figure _____

3. **Study the *Trinidad: San Fernando* map on page 69 of your atlas.**

   (a) State the four-figure grid reference for each of the following locations.

   (i) The Public Transport Service at Bontour Point

   (ii) The Mangrove area in Tarouba Bay

   (iii) The Roundabout at Mon Repos

   (iv) Pointe-à-Pierre

   (v) Tarouba Hill

   (b) State the six-figure grid reference for each of the following Trigonometrical Stations.

   Trigonometrical Stations south of northing 38

   (i) No. 2145 (68)

   (ii) No. 92 (586)

   (iii) No. 2107 (87)

   (iv) No. 2144 (82)

   (v) No. 2143 (63)

Trigonometrical Stations north of northing 38

(vi) No. 2108 (130)

(vii) No. 2163 (13)

(viii) No. 1067

(ix) No. 2166 (25)

(x) No. 2147 (42)

Sketch maps

1 Study the *Grenada* map on page 56 of your atlas. Complete the sketch map showing the grid square B3 by following the instructions below:

(a) Insert:

   (i) St Patrick river

   (ii) Levera Pond

   (iii) Lake Antoine

   (iv) the parish boundary

   (v) Sauteurs (important town)

   (vi) the main road connecting R. Sallee and La Poterie along the coast.

(b) Shade:

   (i) the Atlantic Ocean

   (ii) the land with height of 200 m and above.

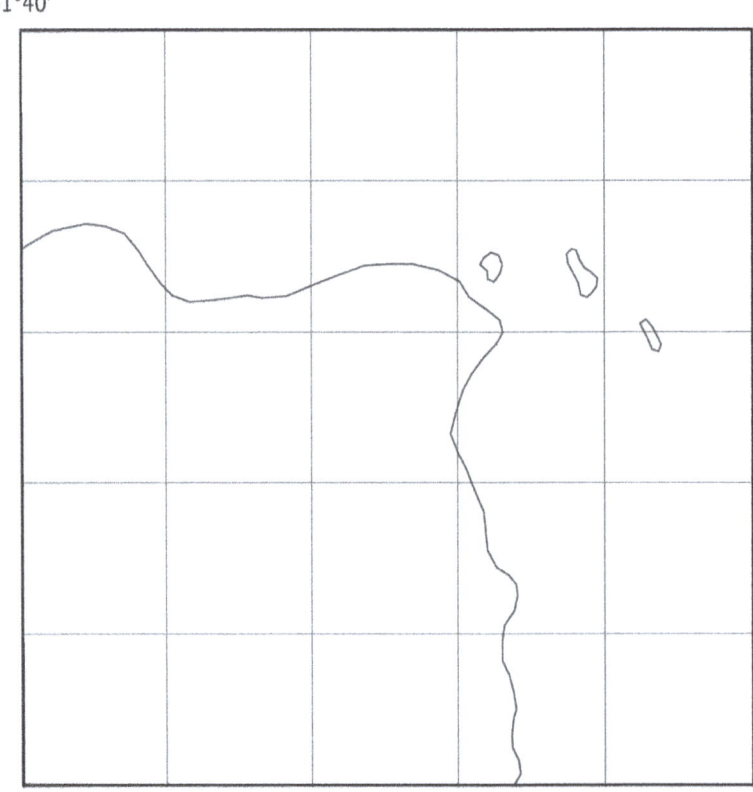

2 Study the *Barbados: Cove Bay* map on page 59 of your atlas. For the area bounded by easting 27 and northings 88 and 89, add detail onto the sketch map following the instructions below.

(a) Insert:

 (i) the 20 m contour line which runs from Gay's Cove (Cove Bay) in the south to where it crosses northing 89 in the north

 (ii) the 70 m contour line which runs from northing 88 (The Graveyard) to the inland cliff at grid reference 276882

 (iii) the depression at grid reference 275881.

(b) Insert the secondary road, surfaced and other roads and tracks in the area.

(c) Insert the inland cliff area from where it crosses northing 88 (277880) in the south to 272886.

(d) Shade the cultivated area.

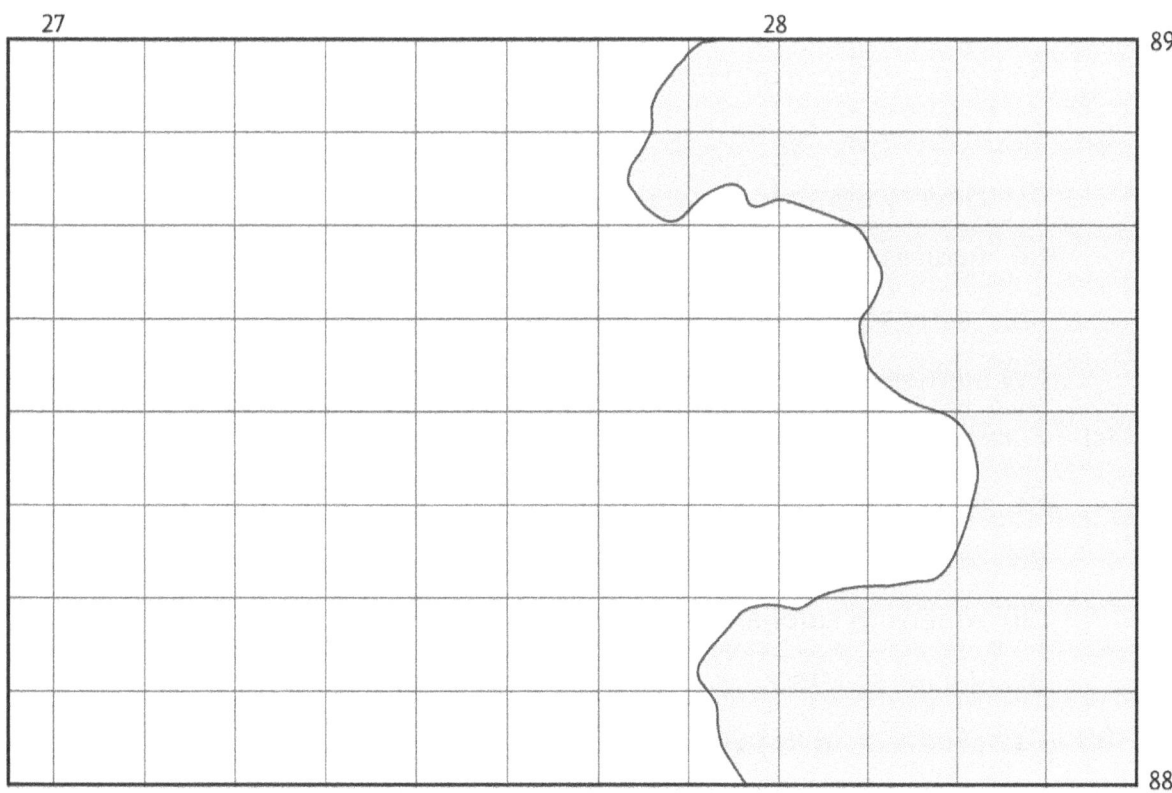

Scale

1. **Study page 9 of your atlas to review key points on the use of map scales and measuring straight-line distances.**

   Calculate the straight-line distance between the following capital cities, by using the maps and their corresponding scales indicated on the pages below. Answers must be written in km.

   (a) *Caribbean: Political*, atlas pages 20 and 21

       (i) Havana and Kingston _____

       (ii) Havana and Nassau _____

       (iii) Georgetown and Paramaribo _____

       (iv) Bridgetown and Castries _____

       (v) Basseterre and St George's _____

       (vi) Roseau and Kingstown _____

       (vii) Port-au-Prince and Port of Spain _____

   (b) *South America: Political*, atlas page 89

       (i) Lima to Bogotá _____

       (ii) Caracas to Montevideo _____

       (iii) La Paz to Brasília _____

       (iv) Santiago to Buenos Aires _____

   (c) *Europe: Political*, atlas page 97

       (i) Dublin to London _____

       (ii) Oslo to Kyiv (previously Kiev) _____

       (iii) Athens to Ljubljana _____

       (iv) Vilnius to Minsk _____

   (d) *Asia: Political*, atlas page 117

       (i) Beijing to P'yŏngyang _____

       (ii) Taipei to Tokyo _____

       (iii) Islamabad to Manila _____

       (iv) Baghdad to Dili _____

2 **Study page 9 of your atlas to review key points on the use of map scales and measuring curved distances.**

Calculate the following distances (in km) by using the maps on the pages below and their corresponding scales.

(a) *Dominica*, atlas page 52

    (i) The western border of St Patrick _____

    (ii) The northern border of St Paul _____

    (iii) The southeastern border of St Andrew _____

    (iv) The shared border between St John and St Andrew _____

(b) *Antigua and Barbuda*, atlas page 50

    (i) Along the main road from Bolans to Urlings (Antigua) _____

    (ii) Along the main road from Five Islands Village to St John's (Antigua) _____

    (iii) Along the main road from Parham to Pigotts (Antigua) _____

    (iv) Along the track from Dulcina to Cocoa Point (Barbuda) _____

(c) *Northern Africa*, atlas pages 114 and 115

    (i) Along the Nile river from Wadi Halfa, Sudan to Merowe, Sudan _____

    (ii) Along the Congo river from the Kisangani airport, Democratic Republic of the Congo to the Kalemie, on the banks of Lake Tanganyika _____

    (iii) Along the road from Tindouf (Algeria), travelling through Abadla and Ksabi to El Goléa _____

    (iv) The international border between Saudi Arabia and Yemen _____

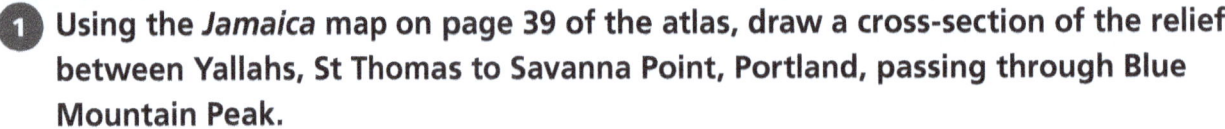

Cross-sections

1 **Using the *Jamaica* map on page 39 of the atlas, draw a cross-section of the relief between Yallahs, St Thomas to Savanna Point, Portland, passing through Blue Mountain Peak.**

Label the lines on the *y*-axis with the correct height in m. Each line (0.5 cm) represents an increase of 200 m in height. Start at 0 m and the final height should be 2400 m.

HINT: Use the key (Relief metres) to see where the height of the land changes, since each colour change shows a change in relief. Be sure to include the point height of Blue Mountain Peak.

2 Using the *St Kitts and Nevis* map on page 49 of the atlas, draw a cross-section of the relief across Nevis from the southern coast through Saddle Hill and Nevis Peak to the northern coast.

The horizontal distance (*x*-axis) should be expanded by a factor of 2, to create a wider base for your cross-section (i.e. it should cover the entire length of the grid provided).

Label the lines on the *y*-axis with the correct height in m. Each line (0.5 cm) represents an increase of 100 m in height. Start at 0 m and the final height should be 1000 m.

HINT: Use the key (Relief metres) to see where the height of the land changes, since each colour change shows a change in relief. Be sure to include the point heights of Saddle Hill and Nevis Peak.

Large scale mapwork (Ordnance survey maps): St Vincent: La Soufrière volcano

Using page 55 of the atlas, study the map of *St Vincent: La Soufrière volcano,* on a scale of 1:50 000, and answer the following questions.

1 Calculate the following distances in km to the nearest 100 m:

(a) the straight-line distance between Porter Point and Goat Point in the northeast corner of the map _____

(b) the length of the Dry River (grid square C10), from its source to its mouth at the Sports Ground (along the coastline). _____

2 Give the 8-point compass direction of Campobello Bay from Baleine Bay. _____

3 Measure the grid bearing of the Police Station (PS) in Sion Hill from the Clinic in Owia. _____

4 Calculate the gradient from the mouth of the river at Tucker Bay to the southeast corner of grid square D8. _____

5 Using map evidence only, identify TWO features of the relief on the map.

_____
_____
_____
_____

6 Using map evidence only, identify TWO features of the drainage on the map and suggest reasons for the observed features.

_____
_____
_____
_____
_____
_____

7 Use map evidence to suggest TWO reasons for the location of settlements and communication on the map.

_____
_____
_____
_____
_____

Barbados: Cove Bay

Using page 59 of the atlas, study the map of *Barbados: Cove Bay*, on a scale of 1:10 000, and answer the following questions.

1 State the direction of:

(a) Pico Teneriffe to Paul's Point _____

(b) Cave Hill to The Risk. _____

2 State the four-figure grid reference of Laycock Bay. _____

3 State the six-figure grid reference of Trigonometrical Station M27. _____

4 Calculate in km to the nearest 100 m:

(a) the straight-line distance between Trigonometrical Station M27 and the bridge at grid reference 278878 _____

(b) the distance along the secondary road (surfaced) from grid reference (274870) to where it crosses the parish boundary. _____

5 Using evidence from the map, describe TWO features of the relief on the map north of northing 88.

_____

_____

_____

_____

6 Use map evidence to describe TWO features of the drainage on the map.

_____

_____

_____

_____

7 Use map evidence to suggest TWO reasons for the distribution of vegetation on the map.

_____

_____

_____

_____

8 You have been asked to develop plans for a hotel. Describe the location you would select for this development and provide ONE reason to support your choice.

Trinidad: Port of Spain

Using page 68 of the atlas, study the map of *Trinidad: Port of Spain*, on a scale of 1:25 000, and answer the following questions.

1 Using map evidence, suggest TWO reasons for the distribution of the built-up area on the map extract.

_____

_____

_____

_____

_____

2 Using map evidence, compare the relief of the areas to the north and south of northing 77.

_____

_____

_____

_____

_____

3 Using map evidence, compare the drainage in the areas to the north and south of northing 77.

_____

_____

_____

_____

_____

4 Explain ONE advantage and ONE disadvantage of the location of the Sewage Plant (6475).

_____
_____
_____
_____
_____

Trinidad: San Fernando

Study page 69 of the atlas.

1. State the direction of:

   (a) Tarouba Hill to Naparima Hill _____

   (b) Guaracara Park to Union Park Race Course & Tracks. _____

2. State the four-figure grid reference of Spring Vale. _____

3. State the six-figure grid reference of the northernmost roundabout on Southern Main Road. _____

4. Calculate in km to the nearest 100 m:

   (a) the straight-line distance from the mouth of Marabella River to the end of the pier at Bagatelle Point

   _____

   (b) the distance along the San Fernando By-pass, from the roundabout at Mon Repos to where the road meets northing 35.

   _____

5. Using map evidence, suggest TWO reasons for the distribution of settlement in the area south of northing 38.

   _____
   _____
   _____
   _____
   _____

6. Using map evidence, suggest TWO reasons for the distribution of sugarcane cultivation on the map.

_____

_____

_____

_____

_____

7. Using map evidence, suggest TWO advantages of the location of the oil refinery (695410).

_____

_____

_____

_____

_____

SECTION 2: NATURAL SYSTEMS
Internal structure of the Earth

1 (a) Draw a simple diagram showing the layers of the Earth in the space provided. Add the following labels: crust, mantle, core.

(b) Explain clearly the difference between an *oceanic* and a *continental* crust.

_____

_____

_____

_____

2 (a) Define the term *plate tectonics*.

_____

_____

(b) Study the diagrams about *Continental drift* on page 144 of your atlas to complete the following exercise.

    (i) Describe the distribution of landmasses and water as depicted in the first diagram (200 million years ago).

_____

_____

(ii) Describe the movement of Laurasia and Gondwanaland 150 million years ago.

_____

_____

(iii) Compare the location of South America in the fourth diagram (50 million years ago) to its current location.

_____

_____

(iv) With the aid of a labelled diagram, briefly describe what causes plates to move.

_____

_____

_____

_____

_____

_____

3 **Study the *World: Tectonics* map on page 144 of your atlas to state the direction and speed for each of the following plates.**

(a) Cocos plate _____

(b) Arabian plate _____

(c) Scotia plate _____

(d) Eurasian plate _____

(e) South American _____

4. With the aid of well-labelled diagrams, use each heading provided below to compare *convergent* and *divergent* plate boundaries. Give specific examples for each.

   (a) Movement of tectonic plates

   _____
   _____
   _____
   _____

   (b) Creation or destruction of the Earth's crust

   _____
   _____
   _____
   _____

5 **Working from page 26 of your atlas, complete the following diagrams.**

(a) Add arrows to both diagrams to show the direction of movement for the plates.

(b) Add the labels 'deep-sea trench' and 'volcanoes' on diagram A.

(c) Add the label 'ocean ridge' and 'magma' to diagram B.

(d) Label the subduction zone with the letter S, on diagram A.

(e) Write the name of each plate boundary represented along the lines provided.

A

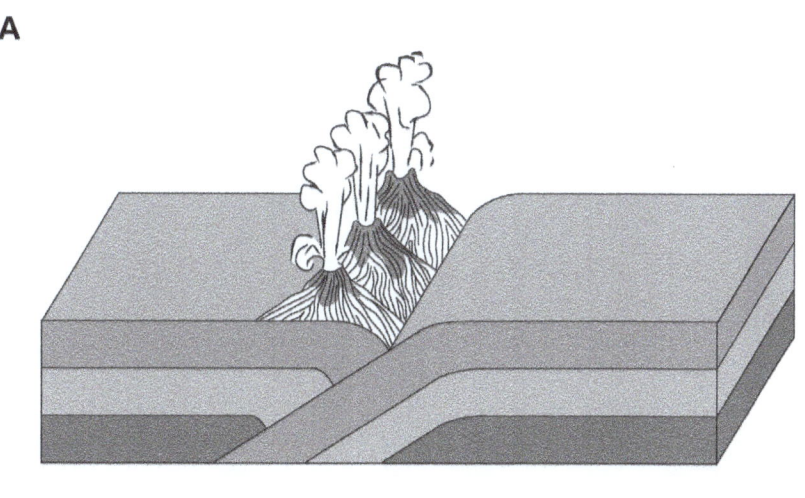

_____

B

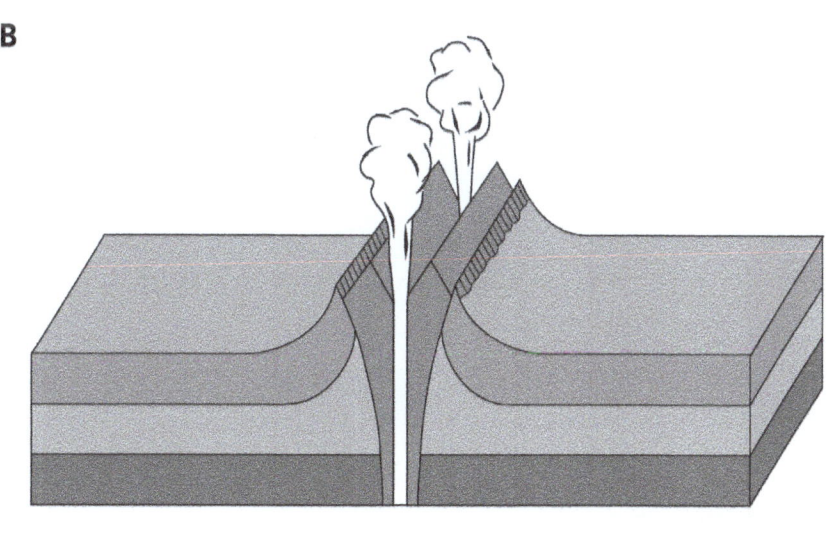

_____

6. (a) Carefully study the *Caribbean: Structure and Geology* maps on page 25 of your atlas. Add the following to the blank map provided:

   (i) the names of the tectonic plates labelled: A, B, C, D, E

   (ii) the names of the two ocean trenches labelled 1 and 2

   (iii) a label for the Lesser Antilles volcanic island arc.

(b) Name island F.

(c) Study the arrows which show the direction of plate movement on the map. Identify TWO areas along the edge of the Caribbean plate boundary with an **X**, where a subduction zone is located.

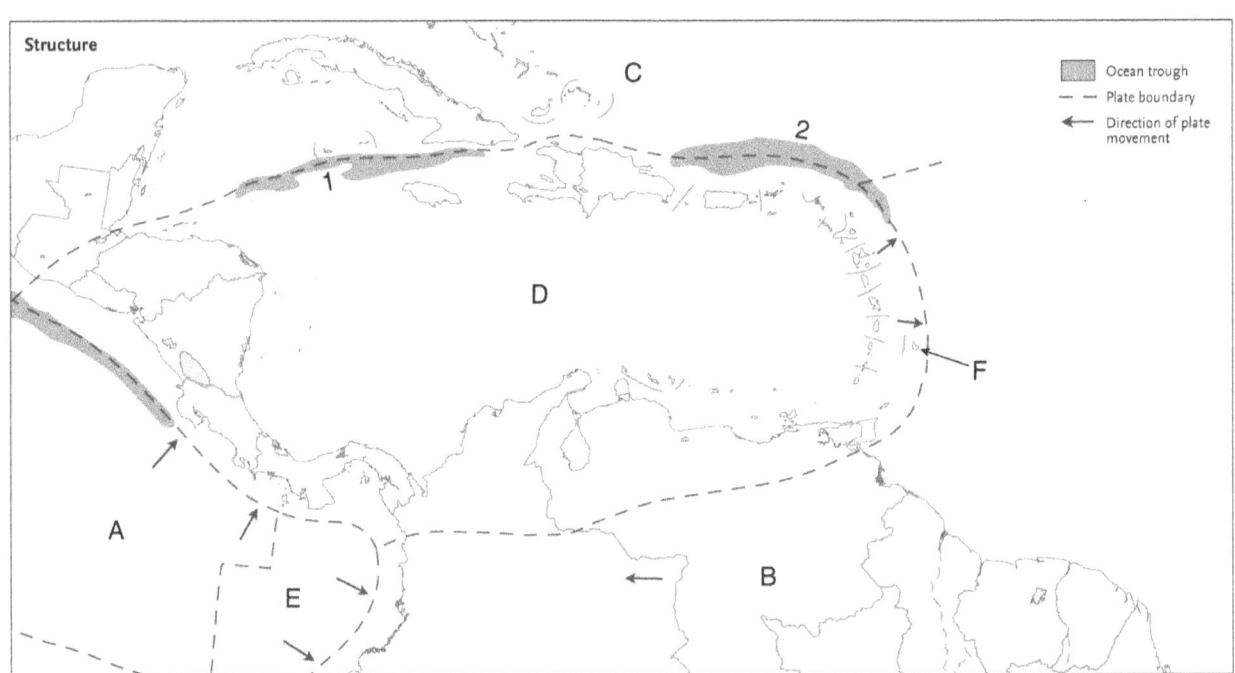

7 With the aid of a well-labelled diagram, describe what happens at a transform plate boundary. Give examples in your response.

8 Using specific examples, explain how fold mountains are formed at collision zones. Use a labelled diagram to aid in your response.

9 Briefly explain why Barbados is not a part of the volcanic island arc in the eastern Caribbean.

Earthquakes: processes

1 (a) Define what an earthquake is.

_____

_____

(b) Describe how earthquakes occur at convergent plate boundaries.

_____

_____

_____

_____

2 Study page 26 of your atlas which shows *Caribbean: Earthquakes* to answer the following questions.

(a) State ONE aspect of earthquakes in the Caribbean region that is being depicted on the map.

_____

(b) Account for the distribution of earthquake epicentres on the map.

_____

_____

3 Complete the diagram below by adding the following labels:

| sea | tectonic plates | focus | epicentre | seismic waves |

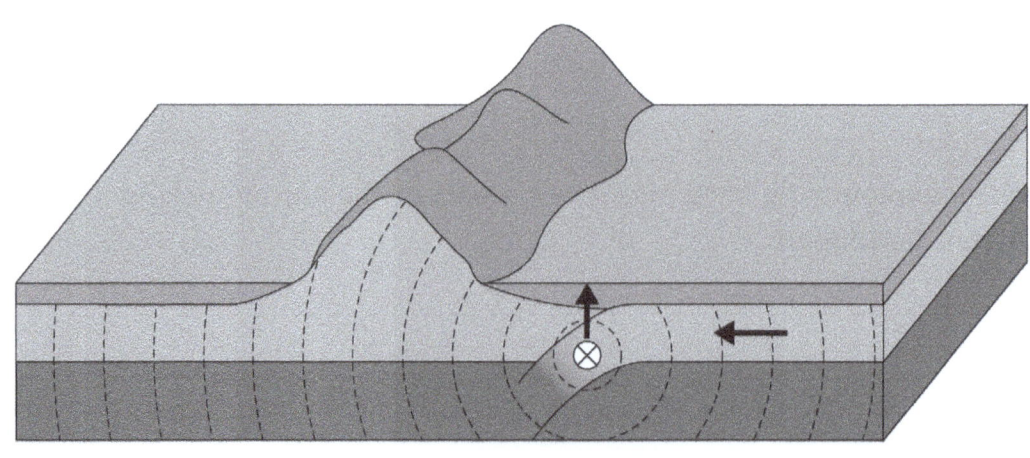

4 State what the Richter scale measures.

___

5 Information on the *Caribbean: Earthquake* map on page 26 of your atlas, indicates that several earthquake epicentres of magnitude 5.0 or greater have been recorded in the Caribbean region since 1900.

Complete the table by inserting the correct information for each category. The first one is done as an example.

NAME OF COUNTRY	YEAR EARTHQUAKE OCCURRED
Jamaica	1907
	2010
Trinidad and Tobago	
	1918
Dominican Republic	
	2007

6 Give ONE reason for the 'frequent strong' earthquakes recorded in the specific areas highlighted on the map on page 26 of your atlas.

___
___
___

Volcanoes: structure and formation

1 (a) Define what a volcano is. _____

_____

(b) Identify the features of a volcano labelled **A–G** in the following diagram.

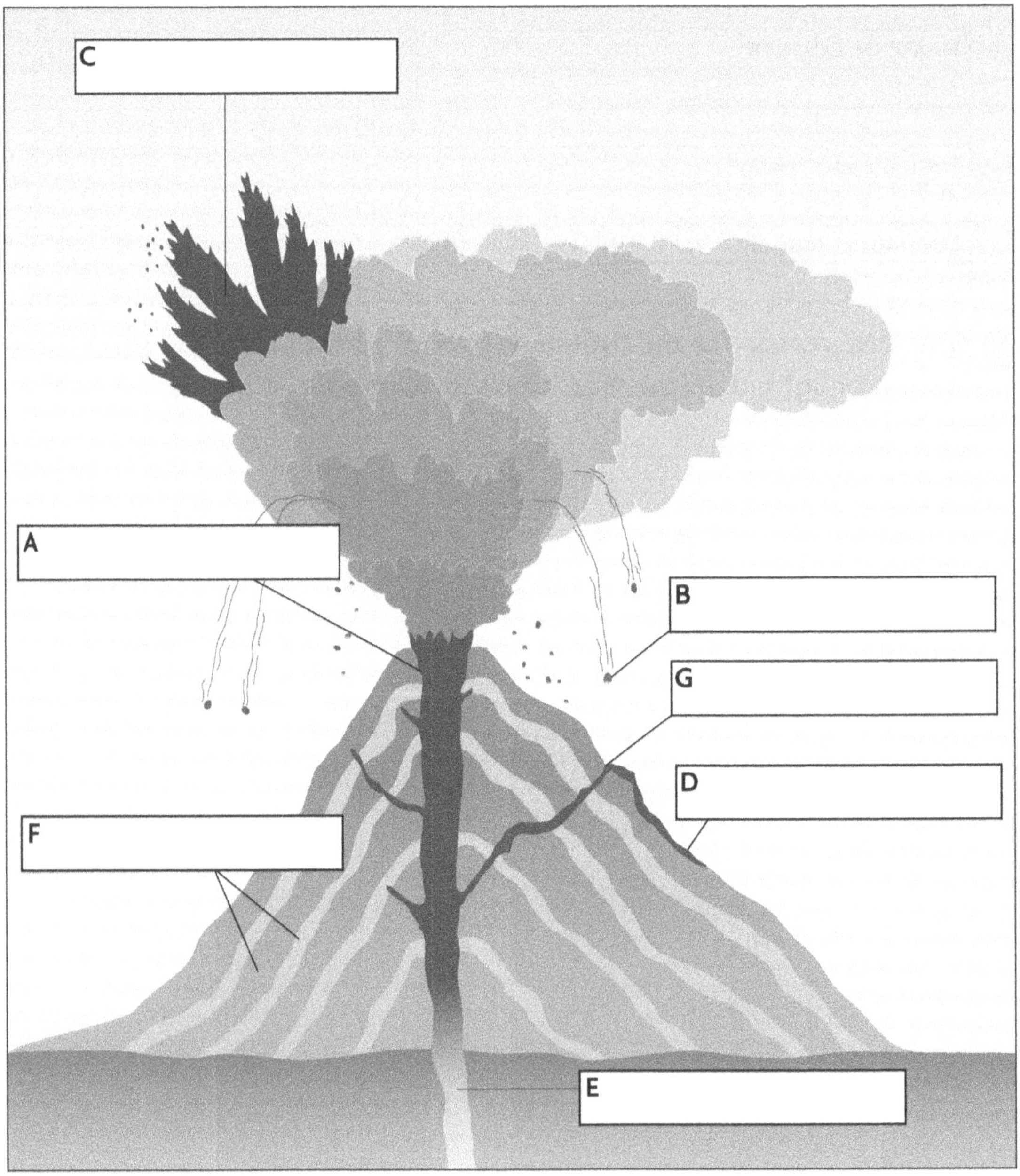

2 (a) What type of volcano is shown in question **1**?

Circle the correct option from those listed below.

   A Composite   B Ash and Cinder Cone   C Basalt Plateau   D Shield

(b) Justify your answer to part **(a)** above.

_____
_____

3 With the aid of a well-labelled diagram, describe how volcanoes are formed at subduction zones.

_____
_____
_____

4 Using specific examples, explain the concept of 'hot-spot' volcanoes.

_____
_____
_____
_____
_____
_____

5 **(a)** Differentiate between 'intrusive' and 'extrusive' volcanic features.

_____

_____

_____

(b) Identify the intrusive volcanic features labelled **A–C** in the diagram below.

A _____   B _____

C _____

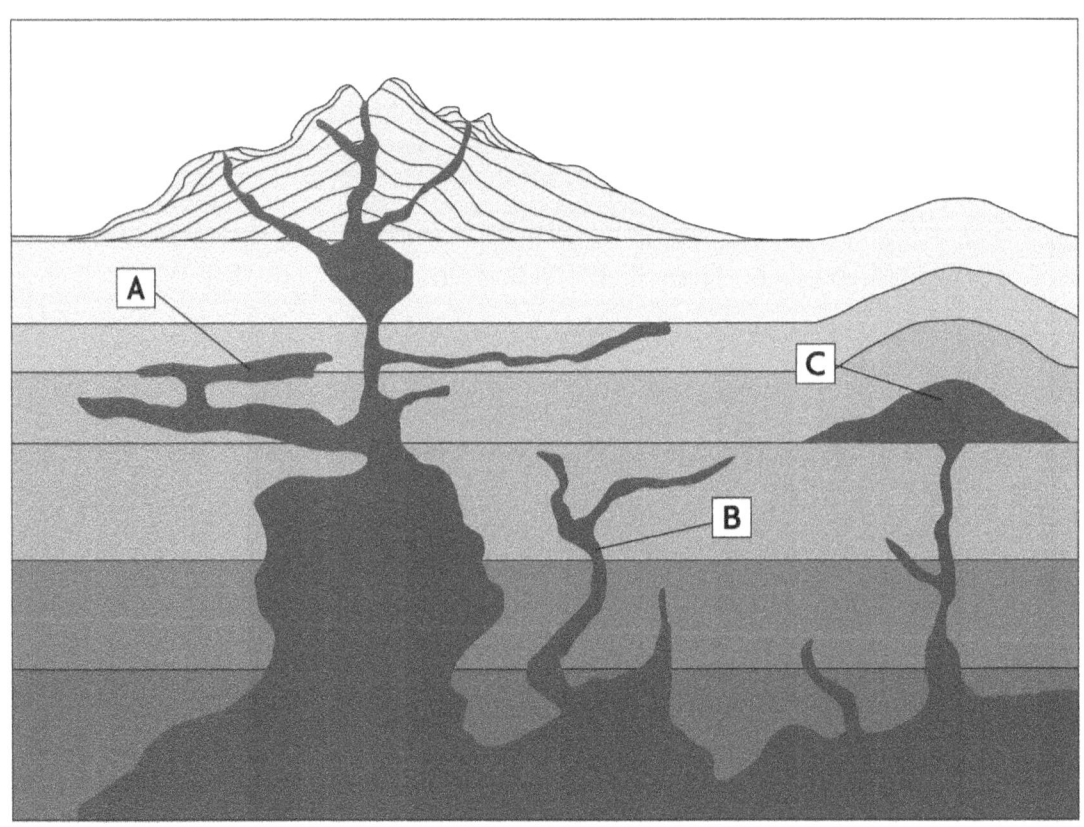

(c) Compare each of the following intrusive volcanic features.

   (i) laccolith and batholith

   _____

   _____

   _____

   _____

   (ii) dyke and sill

   _____

   _____

   _____

   _____

6. How can an intrusive volcanic feature become a part of the surface landscape over time?

_____

_____

_____

7. Draw and label a simple diagram of a shield volcano.

8 Complete the table below to compare the volcano diagram in question 1 with a typical shield volcano.

CHARACTERISTICS	DIAGRAM Q1	SHIELD VOLCANO
Shape	Tall, conical	
Volcanic material	Lava and ash	
Type of lava	Acidic	
Eruptions	Explosive	

9 Explain why basic lava typically forms broad, flat volcanic structures.

_____
_____
_____
_____

10 Explain why plugs in volcanic vents are typically formed by acidic lava.

_____
_____
_____
_____

11 Study the volcanoes table on page 27 of your atlas. State the name of ONE currently active volcano in the eastern Caribbean.

_____

12 Extended Learning

Do some research on calderas and crater lakes to help you draw a series of annotated diagrams showing how a typical volcanic cone may become a caldera with a crater lake.

Volcanoes: volcanic landscapes

1 Briefly explain how the volcanic island arc of the eastern Caribbean was formed.

_____
_____
_____
_____

2 Study the *St Vincent: La Soufrière volcano* map on page 55 of your atlas to answer the following questions.

(a) Discuss how the features of relief, vegetation and drainage on the map have been influenced by the volcanic origins of the island.

_____
_____
_____
_____
_____
_____

(b) The following grid is a section of the northeastern corner of the map. Circle the TWO headlands shown in the map extract provided below.

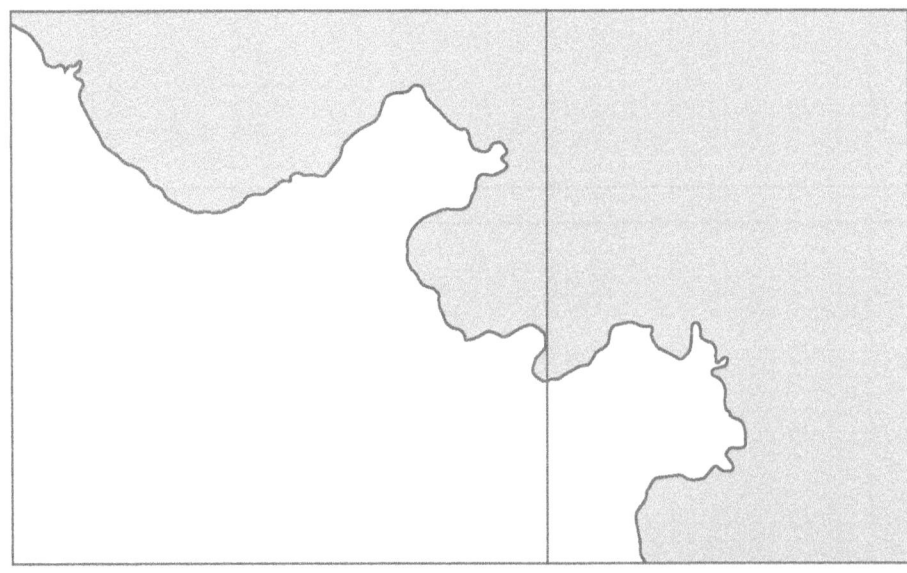

(c) Describe how a past eruption of the La Soufrière volcano may have created these headlands.

_____
_____
_____
_____

3 State ONE way in which the volcanic origin of the eastern Caribbean islands has influenced each of the industries listed.

(a) Agriculture

_____
_____
_____
_____

(b) Tourism

_____
_____
_____
_____

4 Describe TWO ways in which a volcanic eruption may IMMEDIATELY change the physical features of a surrounding landscape.

_____
_____
_____
_____

Weathering

1 Define the following terms:

(a) Weathering

(b) Denudation

2 Identify ONE difference between each of the following pairs of processes.

(a) Weathering and erosion

(b) Mechanical weathering and chemical weathering

(c) Physical weathering and physical biotic weathering

(d) Physical biotic weathering and chemical biotic weathering

(e) Hydrolysis and solution

(f) Frost shattering and exfoliation

3 **Complete the following sentence.**

One similarity between frost shattering and exfoliation is that both processes depend on changes in _____.

4 **Study page 78 of your atlas, *North America: Climate* to answer the questions below.**

(a) Working from the Climate regions map, shade the area on the map provided below which is most likely to experience exfoliation weathering.

(b) Explain your answer to part (a) above.

_____
_____
_____
_____

5 With the aid of a well-labelled diagram, explain the process of frost shattering.

6 Explain the process of carbonation. Use these key terms in your answer:

| carbon dioxide | carbonic acid | calcium carbonate | calcium bicarbonate |

7 How can physical weathering promote chemical weathering processes?

_____

_____

_____

_____

8 **Extended Learning**

Carry out research to answer the following questions.

(a) Define the term *haloclasty*.

_____

_____

(b) Underline the correct word from each bracketed pair to correctly complete the following sentence:

Haloclasty is most likely to occur in (inland/coastal) areas where environmental conditions are (arid/wet).

(c) What is the difference between block disintegration and granular disintegration?

_____

_____

(d) Name ONE type of rock that is likely to be affected by granular disintegration.

_____

Mass movement

1 Define the term *mass movement*.

_____

_____

2 State ONE similarity and ONE difference between landslides and soil creep.

_____

_____

_____

_____

3 Study the map of Guadeloupe on page 51 of your atlas to answer the following questions.

(a) On which island, Basse-Terre or Grande-Terre, would you expect to observe more landslides?

_____

(b) Explain your answer to part (a) above. Use map evidence to support your explanation.

_____

_____

_____

_____

4 Study the *Trinidad: Average annual rainfall* map on page 63 of your atlas.

(a) Based only on information from this map, which location, Icacos or Cumaca, is least likely to experience soil creep?

_____

(b) Explain your answer to part (a). Use map data to support your explanation.

_____
_____
_____
_____
_____

5 The Scotland District in the east of Barbados is made up of clay, shale and sandstone which overlay limestone rocks. This contrasts with the rest of the island, which is made of limestone.
Use an annotated diagram to explain why this difference in geology results in a higher number of mass movement events in the Scotland District when compared to the rest of the island.

_____
_____
_____
_____

6 Discuss how the following human activities can increase the likelihood of landslides.

(a) Building a housing estate at the top of a slope.

_____

_____

_____

_____

(b) Building a train track at the base of a slope.

_____

_____

_____

_____

(c) Rearing grazing animals on a hillside.

_____

_____

_____

_____

7 List THREE signs that a slope is being affected by soil creep.

_____

_____

_____

Limestone: characteristics and processes

1 Study the *Caribbean: Geology* map on page 25 of your atlas. How would you describe the extent and location of limestone in each of the following countries?

The first one is done for you as a guide.

(a) Jamaica

Jamaica is made up almost entirely of limestone; only a small area on the island's eastern side is not limestone.

(b) Puerto Rico

(c) Guadeloupe

(d) Barbados

(e) Trinidad

2 Explain why limestone is permeable. Use these keywords in your answer:

| carbonation | joints |

3 Apart from permeability, state at least TWO other characteristics of limestone.

_____
_____
_____

4 Work from the *Jamaica* map on pages 38 and 39 of your atlas. Compare the drainage density of the Blue Mountains and Manchester parish and write a paragraph explaining the relationship between the drainage density and geology of both areas.

_____
_____
_____
_____
_____
_____

5 What is the difference between the following limestone features?

(a) Depression and doline

_____
_____

(b) Sinkhole and swallow hole

_____
_____

(c) Stalactite and stalagmite

_____
_____

(d) Clints and grykes

_____
_____

6 Draw a labelled diagram to illustrate the following underground karst features:

| cave | stalactite | stalagmite | pillar | underground river |

7 (a) On the map of Jamaica below, shade the area known as 'The Cockpit Country'.

(b) Study the information on The Cockpit Country on page 40 of your atlas to answer the following questions.

    (i) Describe the physical features of The Cockpit Country.

    (ii) Explain how The Cockpit Country was formed.

Weather and climate

1 State ONE similarity and ONE difference between weather and climate.

_____
_____
_____

2 Briefly describe where a rain gauge should be placed when measuring rainfall and why.

_____
_____
_____

3 Consider the *Jamaica* map on page 39 of your atlas and the *Jamaica rainfall* map provided below. Referencing Port Antonio and Kingston, use evidence from both maps to discuss the effect of relief on rainfall.

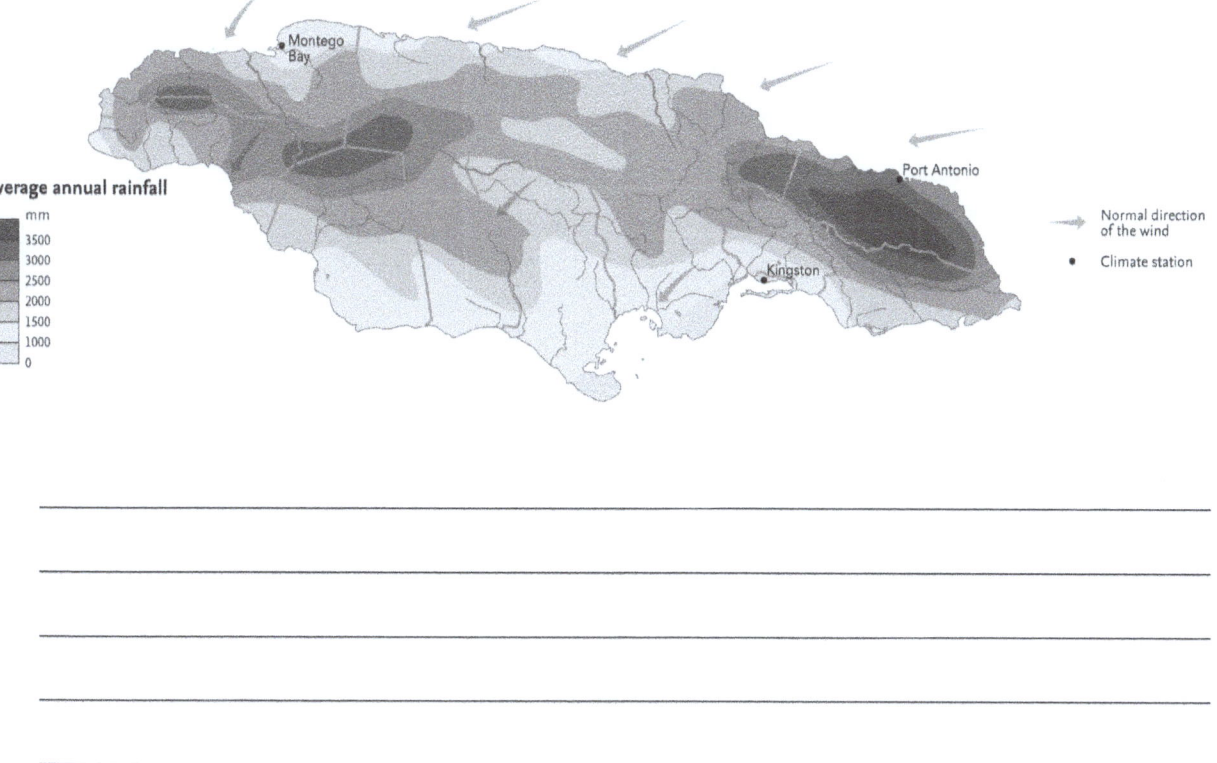

_____
_____
_____
_____
_____

4. Reference and compare the *North America: Climate maps (January and July temperature)* on page 78 of your atlas to discuss how differences in wind direction can affect temperature. Use these key terms in your answer:

| Arctic | tropical | prevailing winds |

5) Study the *Oceania: July temperature* map below. This map shows that Australia's south-eastern edge experiences the country's coolest temperatures. Use evidence from the *Oceania: Relief* map on page 134 of your atlas to discuss how this demonstrates the effect of altitude on temperature.

___

___

___

___

___

___

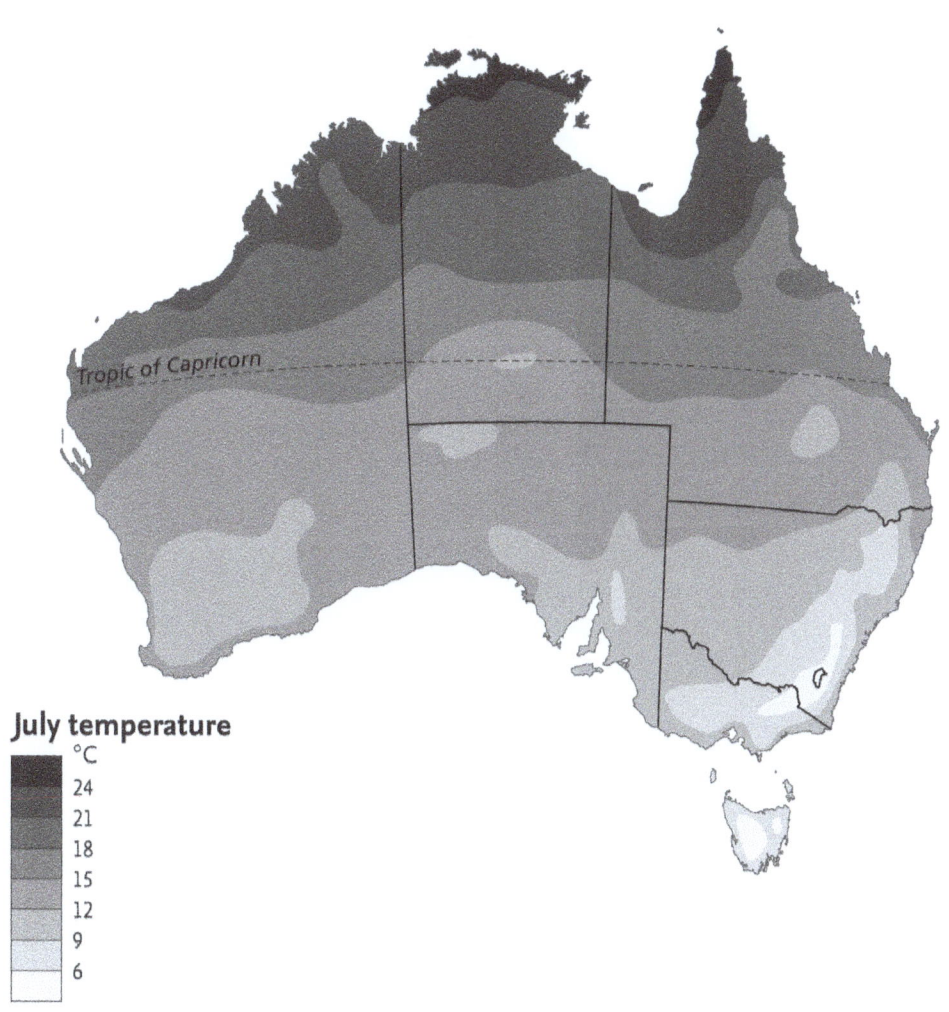

6 (a) Carefully study the *Asia: January temperature* map below.

  (i) State how the map illustrates the relationship between latitude and temperature.

  _____

  _____

  (ii) Explain this relationship by using these key terms in your answer:

  | insolation | atmosphere | distance | absorption |

  _____

  _____

  _____

  _____

  _____

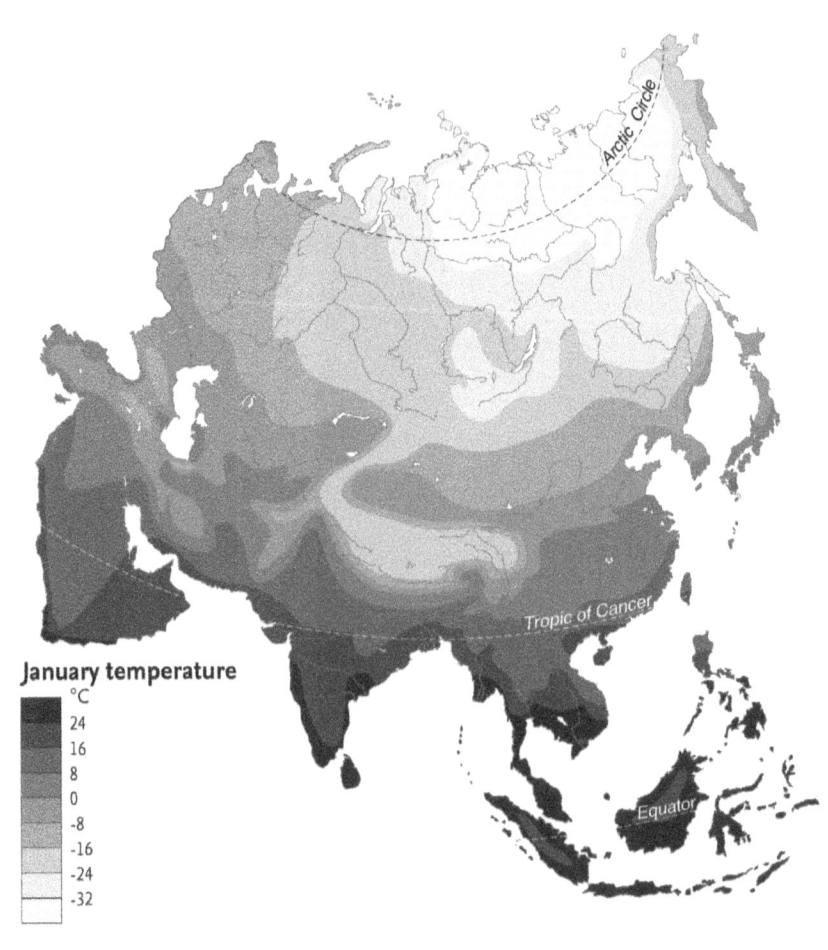

(b) Compare the previous map with the *Asia: July temperature* map on page 118 of your atlas to discuss the effect of the Earth's revolution and seasonality on temperature.

_____

_____

_____

_____

_____

_____

7 Work from pages 22 and 23 of your atlas, which shows *Caribbean: Climate statistics and weather*, to complete the following table. A few examples have been done for you as a guide.

WEATHER STATION	WILLEMSTAD	ROAD TOWN	GEORGETOWN	TEGUCIGALPA	CIENFUEGOS
Climate region			Tropical moist climate		
Wet season	October–January				
Dry season				November–April	
Total annual rainfall					
Annual maximum temperature range					
Average minimum monthly temperature					

8 Georgetown is known to experience two wet seasons. Name the global weather system responsible for this and explain its effect. Use these key terms in your answer:

> frontal   trade winds   overhead sun

_____

_____

_____

_____

9 **Work from the synoptic charts on atlas page 23 to complete the following exercise on Caribbean weather systems.**

(a) Calculate the difference in atmospheric pressure between the hurricane's outermost isobar and the pressure in the hurricane's eye (innermost bar). Show your working.

_____

(b) What category hurricane is shown in the chart?

_____

(c) What is the relationship between the difference in pressure and wind speed?

_____

_____

(d) (i) Study the anticyclone chart and draw the northernmost and southernmost weather symbols.

(ii) Use the key on the page to describe the weather conditions represented by the northernmost and southernmost weather symbols.

WEATHER ELEMENTS/CONDITIONS	NORTHERNMOST	SOUTHERNMOST
Atmospheric pressure (range)		
Temperature		
Wind direction		
Wind speed		
Cloud cover		

(e) With the aid of a well-labelled diagram, describe the formation of cold fronts.

_____

_____

_____

(f) Working from the cold front chart on page 23 of your atlas:

   (i) use the weather symbols to calculate the average temperature to the west and to the east of the cold front. Show your working.

   _____

   _____

   (ii) What does this tell you about the effect of a cold front on weather conditions in the Caribbean?

   _____

   _____

(g) Describe the typical weather conditions that are likely to occur: before, during and after a tropical wave.

_____

_____

_____

Study *Caribbean: Hurricanes* on page 24 of your atlas to answer the following questions.

10 Annotate the diagram below to describe differences in the weather conditions between the eye of the hurricane and the outer bands of the hurricane.

11 (a) Find hurricanes Sandy, Dean, Maria and Dennis on the *Hurricane tracks* map and copy their tracks onto the map below.

(b) State ONE similarity about the direction in which all of the hurricanes travelled.

_____

(c) Identify TWO bodies of water over which the hurricanes formed.

_____

(d) Use examples from the map to identify FOUR locations where a hurricane can die.

_____

_____

_____

_____

(e) State TWO differences between the hurricane tracks of Dean and Sandy.

_____

_____

_____

12 List THREE Caribbean countries that were affected by Hurricane Maria but not affected by Hurricane Dennis.

_____

_____

_____

Global warming and climate change

1 (a) Explain the concept of the greenhouse effect.

_____

_____

_____

(b) Study the *climate change infographic* on page 102 of your atlas. Use the labels provided in the box to complete the diagram below, about the greenhouse effect.

| sun   insolation   cloud   greenhouse gases   ice   ocean   city |

(c) Add the letters below to your diagram.

  (i) Add **S** above the arrow which shows insolation reflected out to space by icy surfaces.

  (ii) Add **C** above the arrow which shows insolation reflected out to space by clouds.

(iii) Add **O** above the arrow which shows insolation reflected out to space by the surface of the ocean.

(iv) Add **A** above each arrow which shows insolation reflected out to space by the atmosphere.

2 Define the following terms:

(a) Greenhouse gas

_____

_____

(b) Global warming

_____

_____

(c) Climate change

_____

_____

(d) Insolation

_____

_____

(e) Radiation

_____

_____

3 List THREE greenhouse gases and for each, state ONE human activity which increases its concentration in the atmosphere.

_____

_____

_____

_____

4 Although methane is a more powerful greenhouse gas than carbon dioxide, why do most discussions about global warming focus on the level of carbon dioxide in the atmosphere?

_____

_____

5 Complete each paragraph below with the correct terms provided in the box below it. The paragraphs are related to the evidence of climate change.

(a) As climate changes the _____ are becoming warmer. This is noteworthy because seas heat up more slowly than _____. Therefore, for the seas to _____ enough heat to raise their _____ the amount of heat being stored around the _____ must be _____.

| Earth | continents | absorb | temperature | significant | oceans |

(b) Ice sheets are also called continental _____. _____ monitor the size of ice sheets, since higher temperatures melt ice. Therefore, the rate at which ice sheets thaw or freeze _____ a change in climate. Additionally, ice sheets also function to _____ sunlight back out into space. Hence, as ice sheets shrink, more sunlight is absorbed at the Earth's _____ and global temperatures can _____ even more.

| reflect | increase | indicates | glaciers | environmentalists | surface |

(c) The ocean is a natural _____ because it absorbs carbon dioxide (CO_2) from the atmosphere. However, when CO_2 _____ in water it creates a chemical called carbonic acid. This means that as more CO_2 is released into the atmosphere by _____ activities the ocean becomes more _____. This is harmful to _____ life since the usual sea _____ are changing and these animals are unable to adjust.

| acidic | carbon sink | dissolves | marine | conditions | human |

6 Study the *Number of Atlantic hurricanes, 1910–2010* graph on page 24 of your atlas to answer the following questions.

(a) Which year recorded the highest number of hurricanes?
_____

(b) In which year was the lowest number of hurricanes recorded?
_____

(c) Which years recorded approximately nine hurricanes?
_____

(d) Describe the general trend of the graph.
_____

(e) Explain how this trend could be a result of climate change.
_____
_____
_____

(f) Describe ONE way this trend could impact Caribbean economies.
_____
_____

7 Study the graph showing changes in coral cover in the Caribbean on page 33 of your atlas to answer the following questions.

(a) Which island experienced a net change of over –60%?
_____

(b) During what years did Port Royal, Jamaica, experience a –20% net change on coral cover?
_____

(c) Describe the general trend of the graph.
_____

(d) Explain how this trend could be a result of climate change.

_____

_____

_____

(e) Describe ONE way this trend could impact marine biodiversity.

_____

_____

8 Study the map of *Relative sea level change, 1960–2015 in the USA* on page 79 of your atlas to answer the following questions.

(a) What has been the estimated sea level change recorded in Florida, between 1960 and 2015?

_____

(b) Describe the general trend in the changes in sea level along the eastern coast of the USA.

_____

_____

(c) Explain how this trend could be the result of climate change.

_____

_____

_____

(d) Describe ONE way this trend could impact coastal settlement on the east coast.

_____

_____

9 Compare the consequences of climate change in the Caribbean with those in EITHER the USA OR the UK, under the following headings:

(a) Sea level rise

_____

(b) Changes in weather patterns

_____

(c) Human health

_____

10 Compare measures to reduce the effects of climate change in the Caribbean with those used in EITHER the USA OR the UK, under the following headings:

(a) Reduced emissions

(b) Sustainable forestry

(c) Education

Ecosystems

1 (a) Define the term *ecosystem*.

2 Describe the importance of the following components in an ecosystem.

   (a) Primary producers

   (b) Decomposers

   (c) Consumers

   (d) Insolation

   (e) Precipitation

   (f) Weathering

3 (a) Use the information in the table to construct a simple food web.

SPECIES	CONSUMED BY
seaweed	rock crab
rock crab	octopus
seaweed	snail
snail	pufferfish
snail	octopus
octopus	conger eel

(b) Name the producer in the food web.

_____

(c) Name the herbivores in the food web.

_____

(d) Name the tertiary consumer in the food web.

_____

(e) List the predators in the food web.

_____

(f) Explain how removing the octopus from the food web could affect:

(i) the pufferfish population

_____

_____

(ii) the conger eel population.

_____

_____

4 (a) How do long, deep roots help grasses in tropical grasslands survive drought?

_____

_____

_____

(b) What advantage do the downward sloping branches of conifers provide during heavy snowfall?

_____

_____

(c) How do deciduous trees conserve water during the dry season?

_____

_____

(d) Why do coastal grasses tend to grow by extending horizontal rooting systems instead of deep roots?

_____

_____

(e) What is ONE advantage a tree may gain by producing tasty, brightly coloured fruit?

_____

_____

(f) Identify TWO species of plants found in the Caribbean that were introduced by humans.

_____

5 Study the *World: Biomes* map on pages 152 and 153 of your atlas. Follow the instructions to complete the map of North America provided below.

(a) Shade an area of: tropical forest, coniferous forest, savanna grassland and desert. Add an appropriate key to identify the shaded areas.

(b) Cross-reference the *World: Climatic Regions* map on pages 148 and 149 of your atlas to label each of the areas you have shaded in (a) with its representative climate.

(c) For each biome identified in (a), add an annotation box to the map to describe ONE adaptation of the vegetation to the climate in that region.

6 **Complete the following diagram of a typical tropical rainforest structure by naming the four layers.**

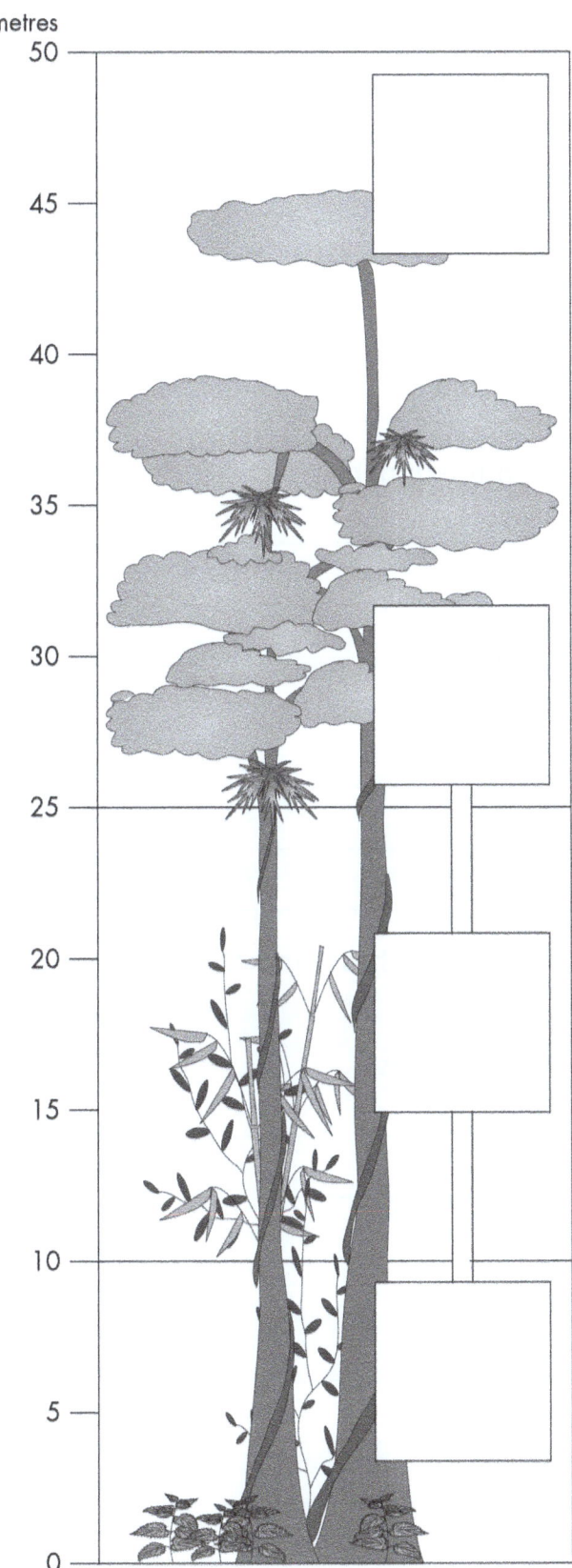

7 With reference to a tropical rainforest biome, explain why:

(a) emergent trees and canopy trees have buttress roots

_____
_____
_____

(b) trees are typically evergreen

_____
_____
_____

(c) vines are more likely to be found in the understorey

_____
_____
_____

(d) there is little vegetation on the forest floor.

_____
_____
_____

8 The tropical rainforest biome has one of the highest biomasses of any ecosystem on Earth.

(a) In your own words, explain what this means.

_____
_____
_____

(b) Provide ONE reason why the biomass of the tropical rainforest is so high.

_____
_____
_____

9 Draw a simple diagram to show how nutrients flow through the tropical rainforest.

10 The tropical rainforest has a high 'nutrient turnover rate'. Explain what this means and why this results in poor-quality soils in the tropical rainforest.

_____
_____
_____
_____
_____
_____

11 The tropical rainforest experiences heavy rainfall daily. Describe how the forest vegetation contributes to this occurrence.

_____
_____
_____
_____
_____
_____

12 Describe the positive and negative impacts of human activities on the biomes of tropical forests in the Caribbean under the following headings.

(a) Positive impact:

   (i) Sustainable management

   _____
   _____
   _____
   _____
   _____

(b) Negative impact:

   (i) Deforestation

   _____
   _____
   _____
   _____
   _____

   (ii) Soil erosion

   _____
   _____
   _____
   _____
   _____

   (iii) Soil exhaustion

   _____
   _____
   _____
   _____
   _____

13 Work from the *Tobago Features and Resources* maps on page 67 of your atlas to complete the following activity on the outline map provided.

(a) Shade and label the forested areas on the island.

(b) Shade and label the Tobago Forest Reserve.

14 Other than the creation of forest reserves, list TWO more strategies that can be used to protect forest resources.

_____
_____
_____
_____

Soil

1 List the FIVE main components of soil.

2 Name the process that produces the mineral component of soil.

3 Name the type of organisms that produce humus in the soil, and give an example of ONE such organism.

4 State the relationship between each of the following in soil:

(a) Air spaces and soil organisms

(b) Clay content and water retention

5 Use the diagram which shows the soil profile of the typical latosol soil layer, to answer the following questions.

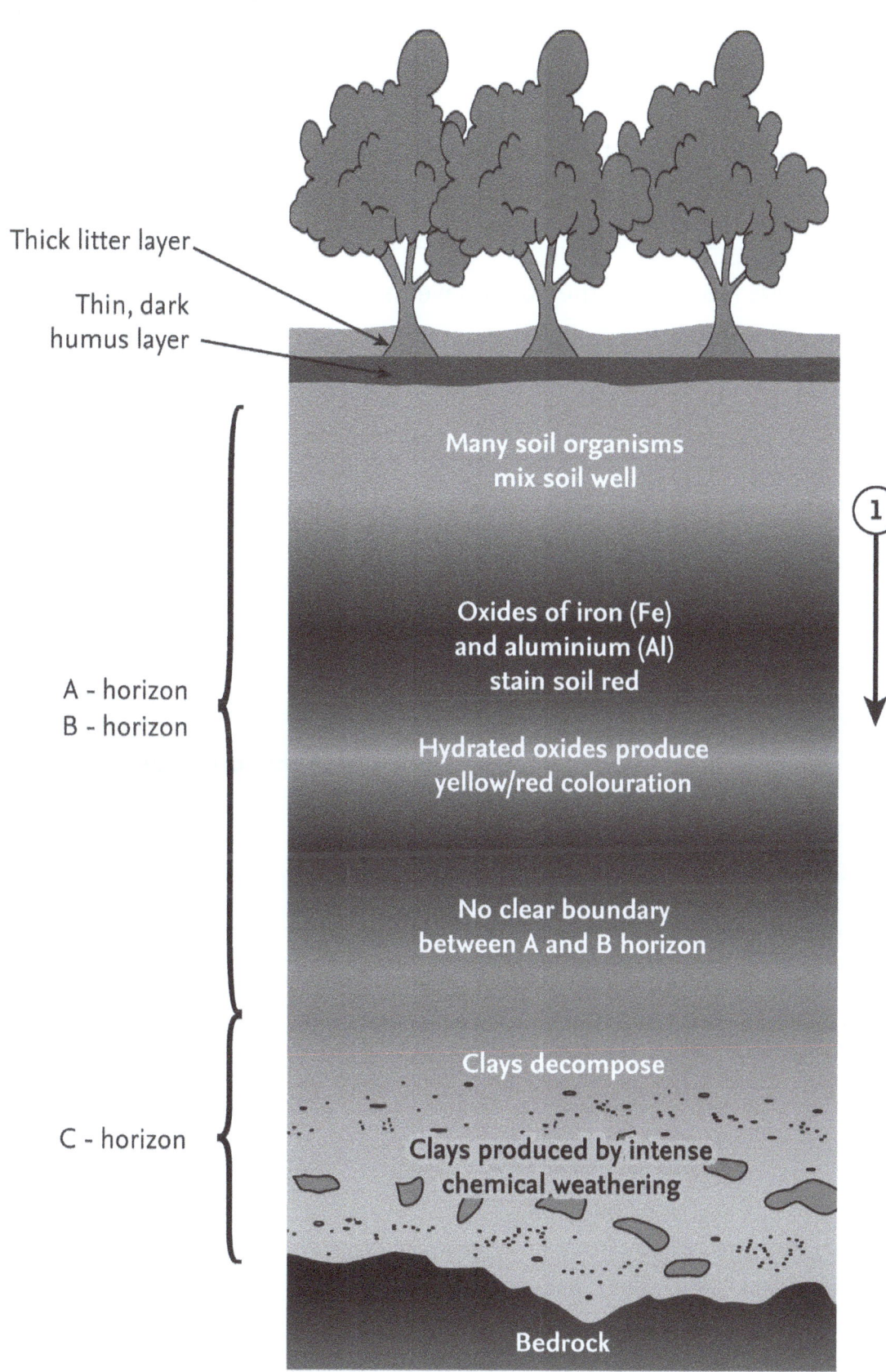

(a) Define the term *soil profile*.

(b) What is the name of the process taking place at the number ① in the diagram?

(c) Give ONE reason why the humus layer is so thin and dark.

6 With reference to latosols, describe the effect of rainfall on soil nutrients.

7 What role does biota (plant and animal life) play in the formation of latosols?

8 Name and describe TWO soil conservation methods.

9 **Study the map of Soil Erosion on page 153 of your atlas.**

   (a) Describe the severity of soil erosion in Haiti.

   _____

   _____

   (b) What is the relationship between the observation in **(a)** and the level of deforestation in Haiti?

   _____

   _____

   _____

   _____

The water cycle and fluvial systems

1 Study the diagram of the hydrological cycle to answer question (a).

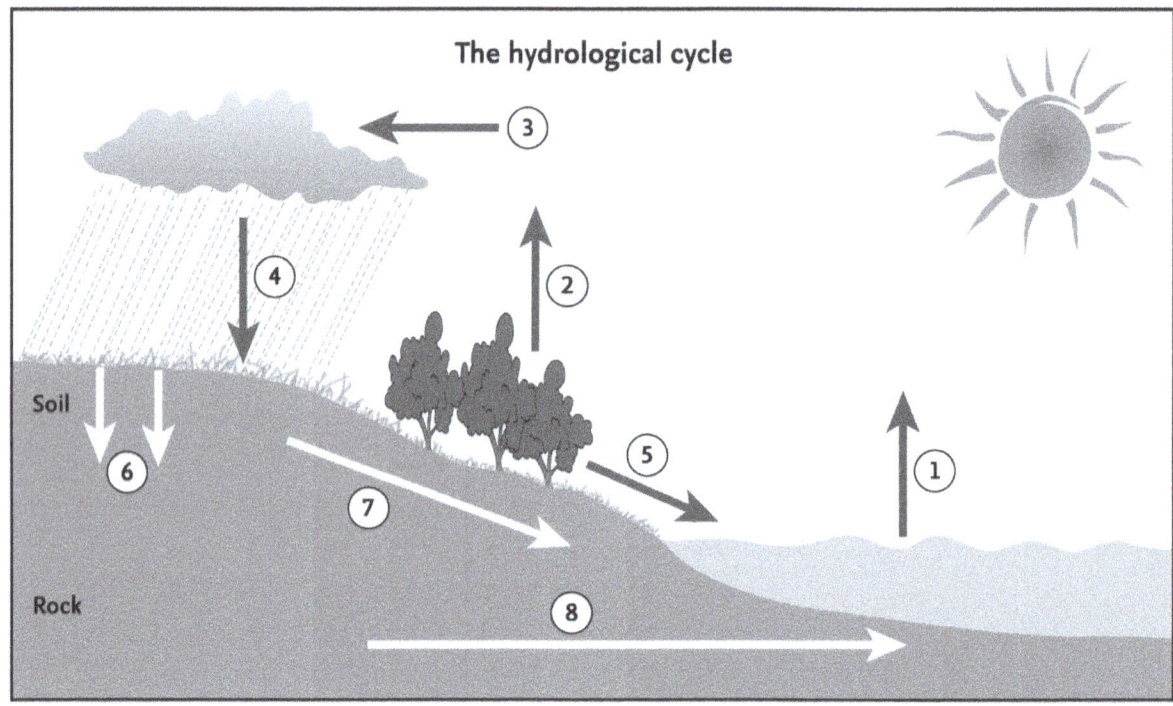

(a) State the names of each process represented by the arrows labelled from 1 to 8.

_____
_____
_____
_____

(b) What is meant by the water table?

_____
_____

2 Study the *Jamaica relief* map on page 38 of your atlas to complete the following activities.

(a) Draw the course of the Black River on the St Elizabeth parish map provided.

(b) Insert the following features:

(i) a tributary

(ii) confluence

(iii) the mouth and source of the river

(iv) one of the river's springs.

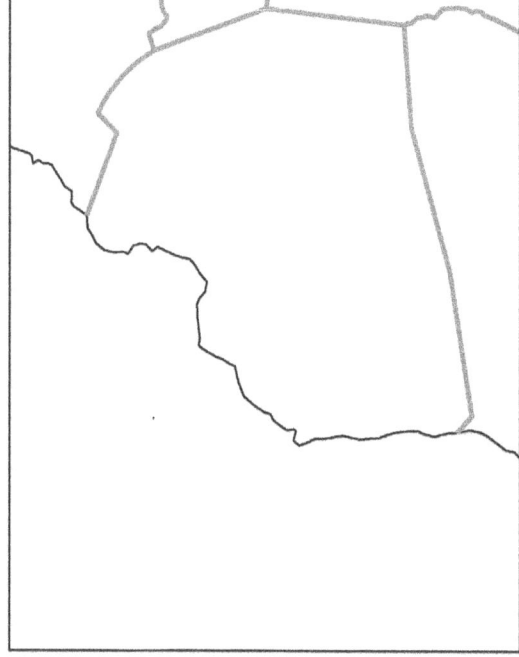

3 Define the following terms:

(a) Watershed

_____

_____

(b) Discharge

_____

_____

(c) Bedload

_____

_____

4 List the THREE types of erosion that occur along the course of a river.

_____

5 Use the diagram showing the long profile of a river to answer the following questions.

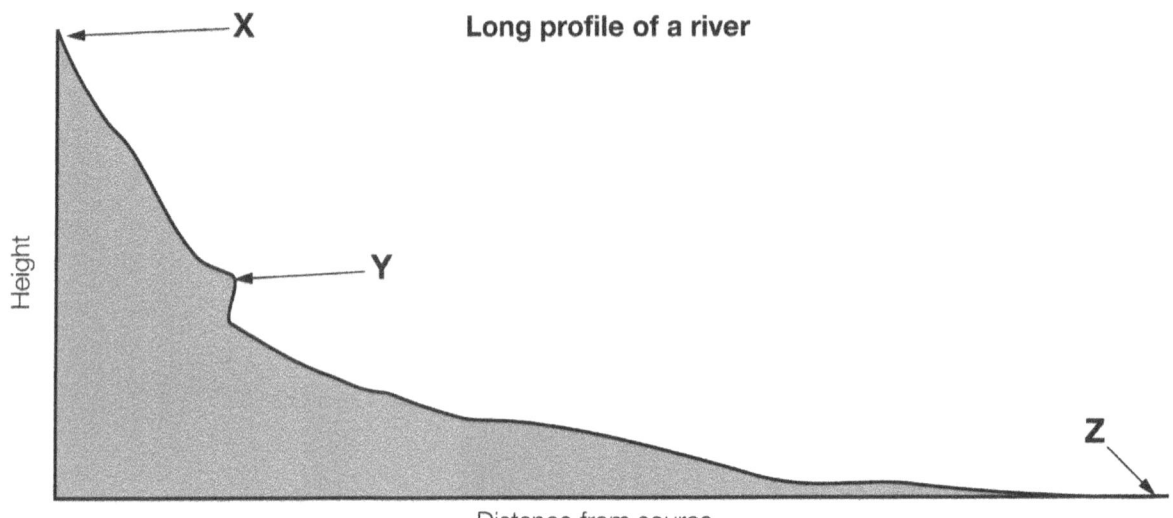

(a) Name the THREE stages of the river labelled **X**, **Y** and **Z**.

_____

(b) Name the feature that is likely to form at point **Y**.

_____

(c) Explain why vertical erosion would be more dominant at point **X** than at point **Z**.

_____

_____

_____

6 With the aid of a well-labelled diagram, explain ONE way a waterfall may be formed along the course of a river.

7 Study the diagram showing the cross-section of a meander to answer the following questions.

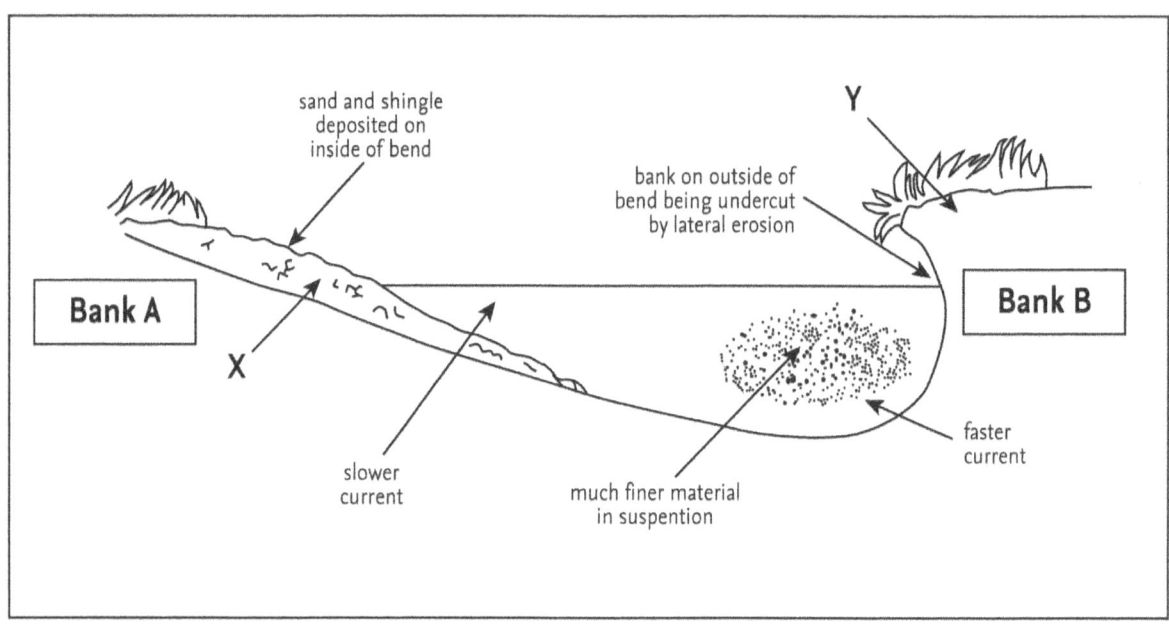

(a) What are the names of the features formed at **X** and **Y**?

_____

_____

(b) On which bank of the river would a concave slope be located?

_____

(c) Which bank of the river would be considered a convex slope?

_____

8 With the aid of a well-labelled diagram, explain the relationship between flooding and levee formation.

9 Draw labelled diagrams to show the stages in the formation of an oxbow lake.

10 Describe ONE reason why braiding may occur along a river's course.

_____

_____

_____

_____

11 Use the diagram showing a river in its lower course to answer the following questions.

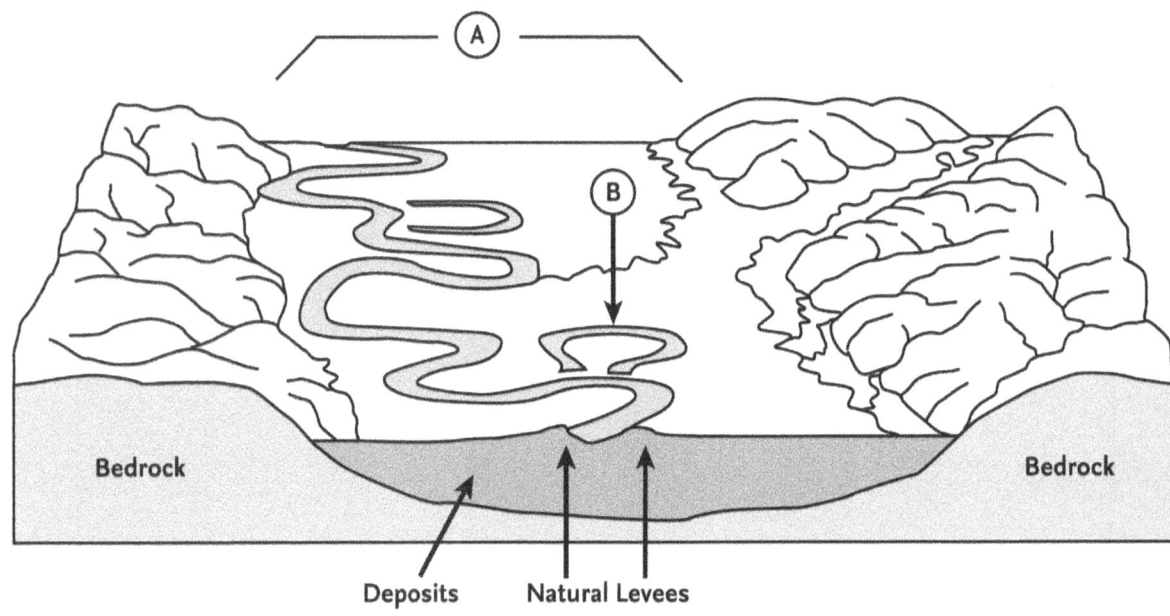

(a) What is the name of the area labelled **A** in the diagram?

_____

(b) Name the featured labelled **B** on the diagram.

_____

(c) Briefly describe how deposition occurs in the lower course of a river.

_____

_____

_____

_____

12 With the aid of a well-labelled diagram, describe the formation of a delta.

___

___

___

___

13 Complete the following table with information about the different types of drainage pattern. An example has been done for you.

DRAINAGE PATTERN	SHAPE	ROCK TYPE	GEOLOGY
Radial			
	Resembles the branches of a tree.		

97

14 Describe TWO ways in which a dam can affect the characteristics of a river valley.

_____
_____
_____
_____

15 Describe how humans have taken advantage of the physical characteristics of floodplains.

_____
_____
_____
_____

Coastal systems

1 Define the following terms:

(a) Wave-refraction

(b) Longshore drift

2 State ONE difference between constructive and destructive waves.

3 (a) Study the aerial photograph of *Bridgetown: Barbados*, on page 57 of the atlas.

(i) Identify the coastal landform that can be seen in Carlisle Bay.

(ii) Which coastal landform is located to the south of Carlisle Bay?

(b) Explain the relationship between these two coastal landforms.

4. (a) Study the *Barbados* map on page 57 of your atlas to complete the map provided.

   (i) Shade and label Carlisle Bay and Needham's Point.

   (ii) Add an arrow to show prevailing wind direction. (Use the *Caribbean Rainfall, winds and currents* map on page 22 of your atlas to assist.)

(b) With reference to your map explain why the west coast is a low-energy coast.

_____
_____
_____
_____

5 With the aid of a well-labelled diagram, describe how a wave-cut platform forms from a retreating cliff. Use these key terms in your answer:

| undercut | hydraulic action | collapse |

6 Use this diagram to answer the following questions.

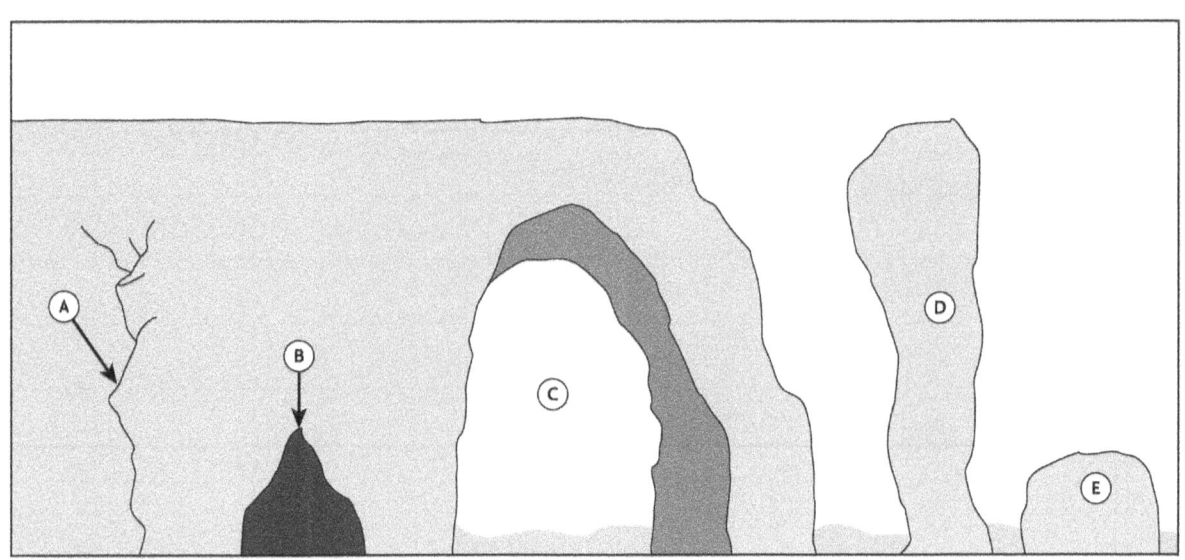

(a) Name the features labelled with the letters **A–E**.

_____

_____

_____

_____

_____

(b) Describe how erosional processes contribute to the formation of the feature at **C**.

_____

_____

_____

_____

_____

7 (a) Work from page 74 of the atlas and use the following instructions to complete the map of Belize provided.

   (i) Draw a circle around the islands that form Belize's barrier reef.

   (ii) Shade and add the name of the group of islands that forms the offshore atoll.

   (iii) Add the location of the Blue Hole Natural Monument. (Use the Features map to help you.)

   (iv) Add the Lighthouse Reef and Glover's Reef.

   (v) Add the locations of fishing ports.

   (vi) Locate and label along the coast: Shipstern Lagoon, Midwinters Lagoon, Northern Lagoon, Southern Lagoon and Placencia Lagoon.

(b) How have changes in sea level resulted in the development of the coastal features identified on your map?

_____

_____

_____

_____

8 **With the aid of a well-labelled diagram, describe the formation of a spit.**

_____

_____

_____

_____

9 Use this diagram of a type of coastal feature to answer the following questions.

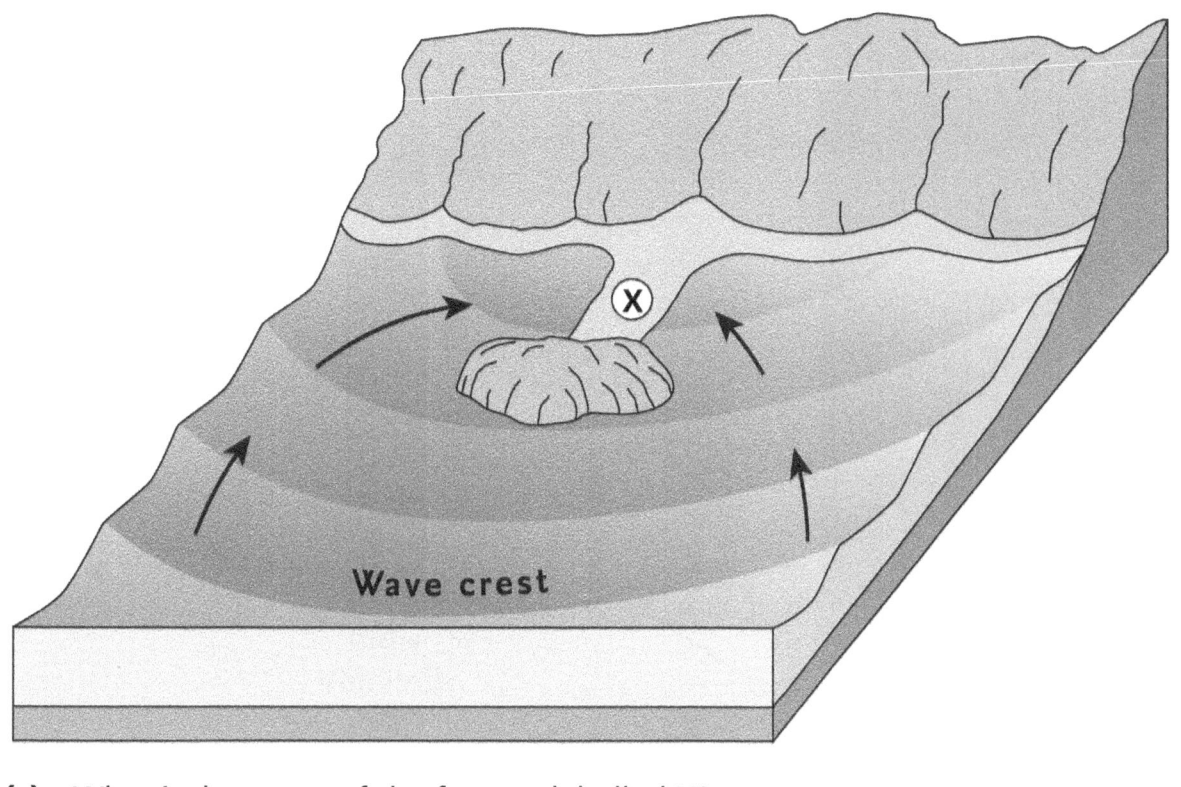

(a) What is the name of the feature labelled **X**?

_____

(b) Describe the formation of the feature in the diagram.

_____

_____

_____

_____

10 Use this diagram of coastal features to answer the following questions.

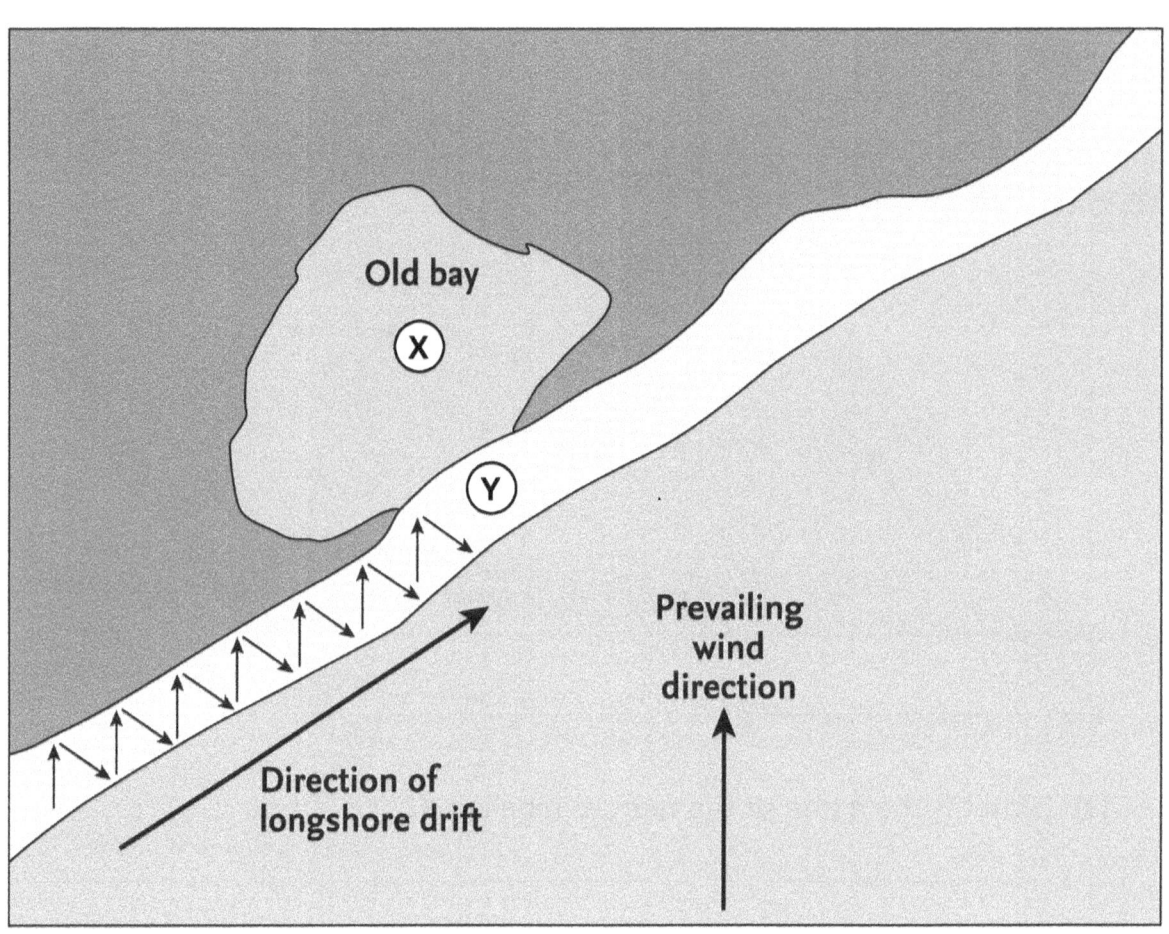

(a) Name the features labelled **X** and **Y** on the diagram.

_____

_____

(b) Describe the formation the feature labelled **Y** on the diagram.

_____

_____

_____

_____

11 (a) Describe what a coral reef is.

_____
_____

(b) List the THREE main types of coral reef.

_____
_____
_____

(c) Describe TWO conditions that are necessary for the successful formation of coral reefs.

_____
_____
_____
_____

12 Study the *Belize* map on page 74 of your atlas to answer the following questions.

(a) How has the barrier reef influenced the formation of coastal lagoons in Belize?

_____
_____
_____

(b) Why is the coral reef system in Belize important to the fishing industry?

_____
_____
_____

(c) Other than fishing, state another benefit the Belizean reef brings to the country's economy.

_____
_____

(d) How could a mass movement event in the Maya Mountains negatively impact the fringing reef on Belize's south coast?

_____
_____
_____
_____

13 **Study the photo below to answer the questions which follow.**

(a) Identify the type of vegetation shown in the photo.

_____

(b) Describe TWO adaptations of this vegetation to semi-aquatic life.

_____
_____
_____
_____

(c) Describe what you see in the photo and explain its importance to offshore fisheries.

_____
_____
_____
_____

14 State TWO ways that human activity can damage coral reefs.

15 (a) Discuss THREE reasons why coral reefs are an important part of the Caribbean's natural heritage.

(b) Identify and assess TWO methods used to conserve coral reefs in a named Caribbean island.

16 (a) What are mangrove wetlands?

_____

_____

(b) Choose ONE option from the list below to describe the importance of mangrove wetlands to Caribbean territories.

   A   Coastal protection

   B   Ecological

   C   Economic benefits

_____

_____

_____

_____

Natural hazards

1 What is the difference between a natural hazard and a natural disaster?

_____

_____

_____

2 (a) Explain why the Caribbean experiences earthquakes and volcanic activity.

_____

_____

_____

(b) Describe TWO hazards associated with earthquakes and volcanoes in the Caribbean.

_____

_____

_____

_____

(c) For a NAMED Caribbean island, discuss the impact of a major earthquake on the physical environment. Include at least THREE impacts.

_____

_____

_____

_____

_____

_____

_____

_____

3 Study the map of an island located in the Caribbean and answer the questions which follow.

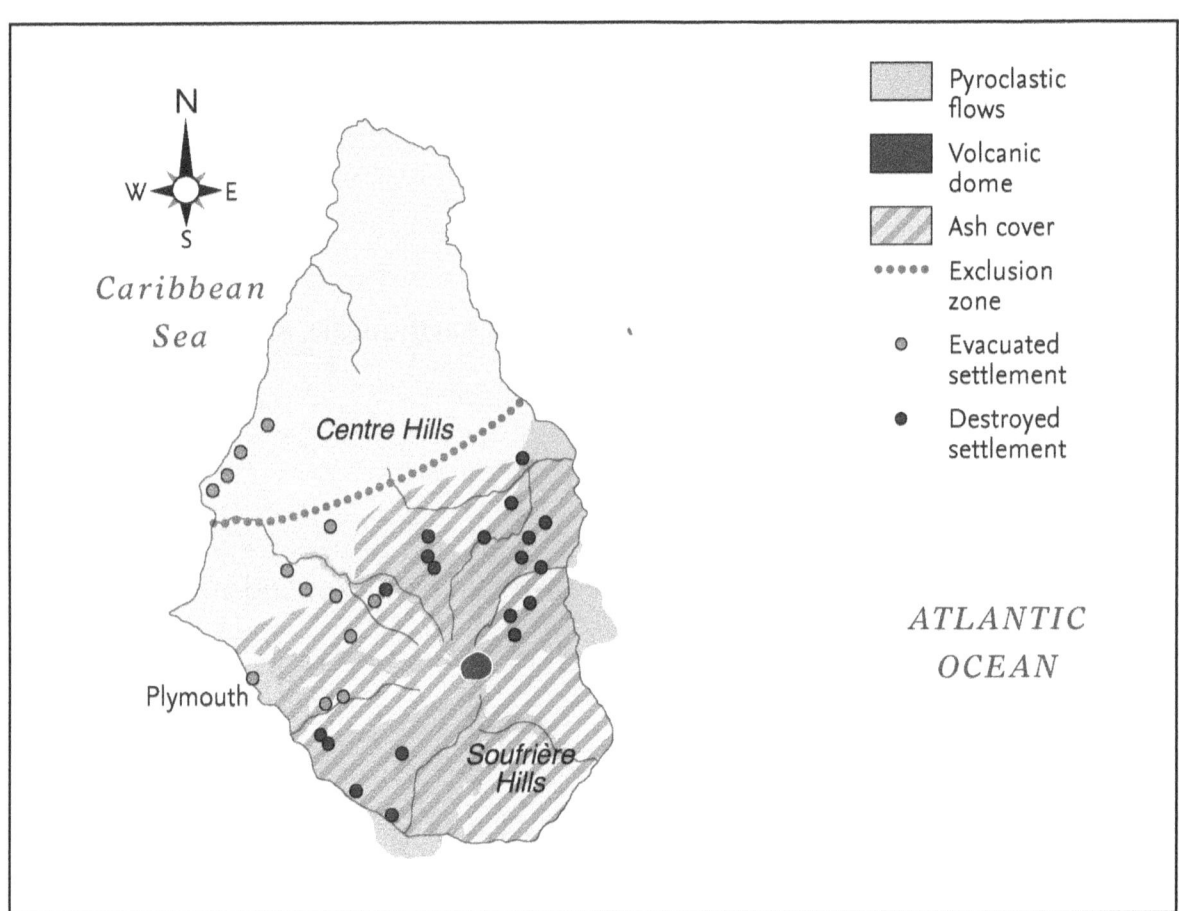

(a) What is the name of the island represented on the map?

_____

(b) Which natural disaster MOST likely affected the island?

_____

(c) What does the dotted line on the map represent?

_____

(d) Which TWO hazards resulted in major damages to the island?

_____

(e) Suggest TWO ways that volcanic activities may be beneficial to this island.

_____

4 (a) Study the following graph showing the number of landslide events in a particular zone.

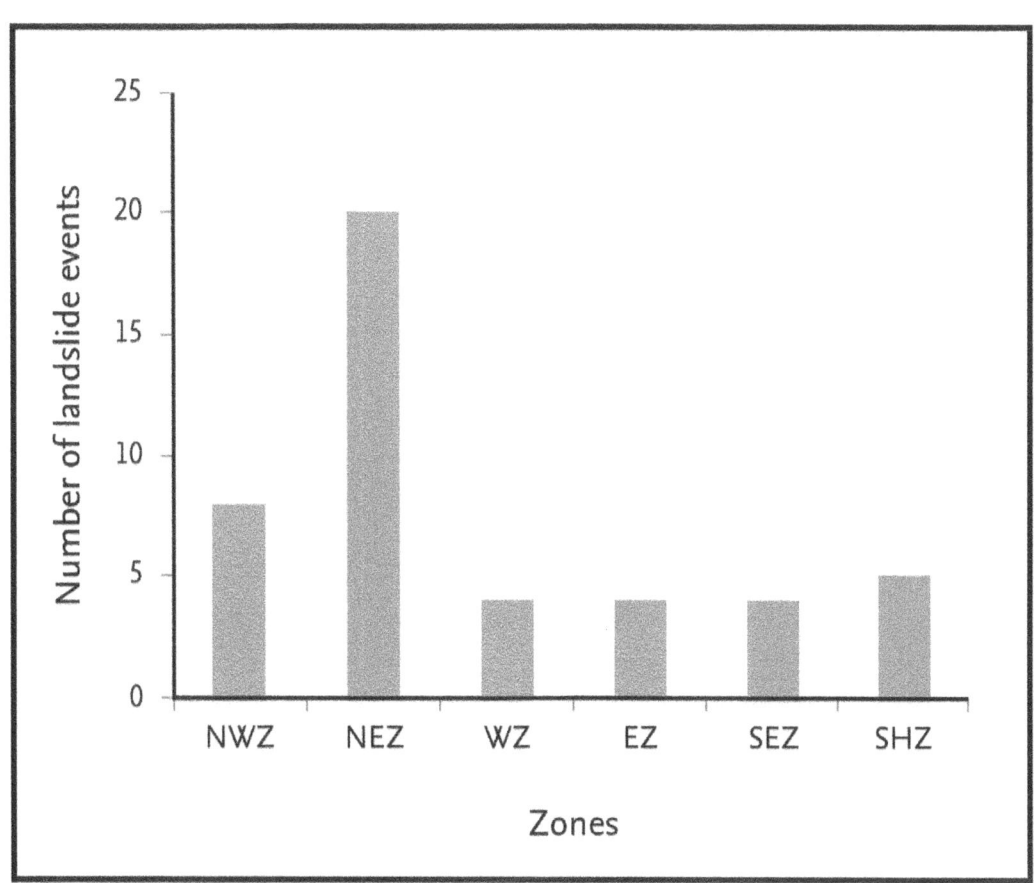

(i) Approximately how many landslides were recorded in zone NEZ?
_____

(ii) How many zones recorded fewer than five landslides?
_____

(iii) Which zone recorded the second highest number of landslides?
_____

(b) In what ways can human activities increase the risk of landslides? State examples in your explanation.
_____
_____
_____
_____

5 Using examples, describe ONE way each flood hazard listed is intensified by human activities in the Caribbean region:

(a) River

(b) Coastal

(c) Flash flood

6 (a) Describe THREE hazards associated with the passage of a hurricane.

(b) When is the hurricane season in the Caribbean?

(c) For a named Caribbean island, explain TWO impacts of a specific hurricane on its economy.

(d) State TWO important things that people should do (i) before, (ii) during and (iii) after a hurricane.

_____

_____

_____

_____

_____

_____

7 Study this diagram of the disaster management cycle to answer the following questions.

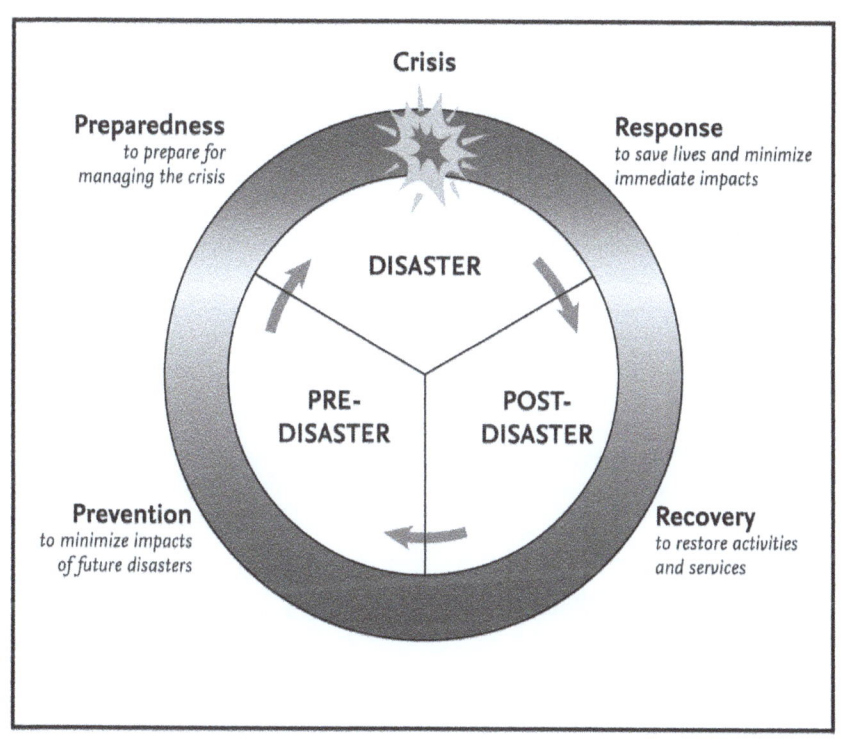

(a) What should be done in the event of an impending disaster?

_____

(b) State TWO types of response that should be applied immediately after a disaster.

_____

_____

(c) Give an example of a long-term recovery approach to natural disasters.

_____

8 (a) What does the acronym CDEMA mean?

_____

(b) What role does CDEMA play in disaster management in the Caribbean region?

_____

_____

_____

_____

9 Describe the role of a national disaster agency in disaster preparedness and management for a NAMED Caribbean island.

_____

_____

_____

_____

_____

_____

_____

SECTION 3: HUMAN SYSTEMS
Population distribution

1. **Study the *World: Population* maps on pages 156 and 157 of your atlas. Answer the following questions.**

   (a) In which hemisphere is most of the Earth's population located?

   _____

   (b) Suggest ONE reason for your answer in **1(a)**.

   _____

2. **Study the *Largest countries by population, 2015* information box on page 156 of your atlas. Answer the following questions.**

   (a) Name the continent with the highest population.

   _____

   (b) Name the TWO largest African countries by population in 2015.

   _____

   (c) State the difference in population size between China and India in 2015.

   _____

3. **Study the *World population growth 1900–2050* graph on page 157 of your atlas. Answer the following questions.**

   (a) Describe the general trend in world population growth from 1900 to 2020.

   _____

   (b) Give the projected population for the year 2050, to the nearest billion.

   _____

   (c) State how population growth from 1960 to 2010 in Latin America and the Caribbean compares to population growth in:

   (i) Europe and Central Asia

   _____

   (ii) sub-Saharan Africa.

   _____

117

4 Reference the *World: Climatic Regions* map on pages 148 and 149 of your atlas and study the maps provided below to discuss the influence of climate on population distribution in:

(a) Africa

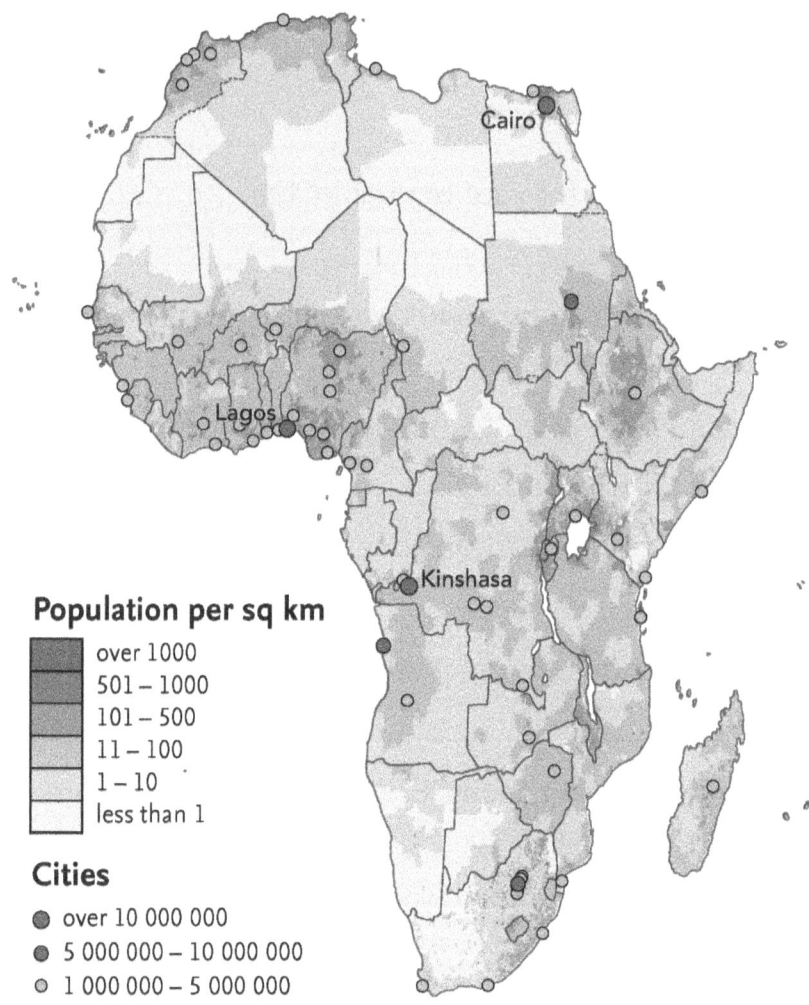

_____
_____
_____
_____
_____
_____
_____

(b) Canada

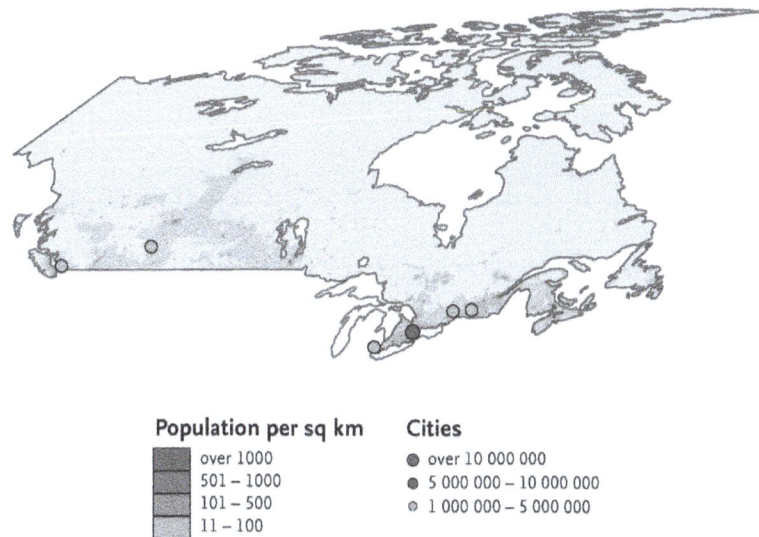

_____
_____
_____

5 Study the *China: Population density* map provided below.

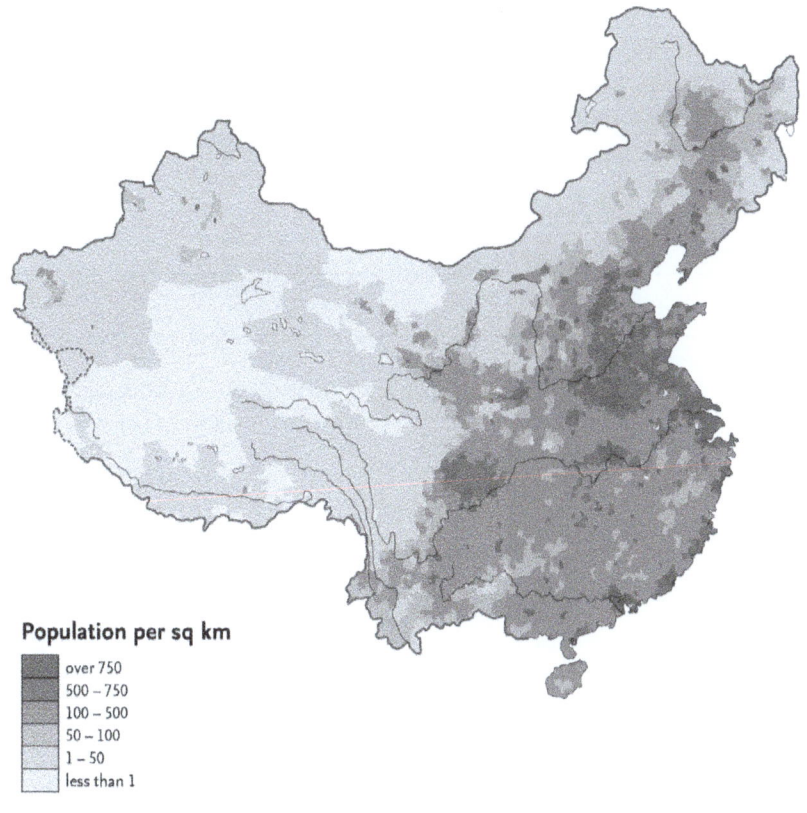

Reference the *China: Relief* map on pages 126 and 127 of your atlas to account for the differences in population density observed between the eastern and western regions.

6. Discuss how access to the following resources can influence population distribution.

   (a) Water

   (b) Fossil fuels

Caribbean population distribution

1. **Draw simple diagrams to illustrate the following settlement patterns:**

   (a) Nucleated

   (b) Linear

   (c) Scattered

   (a)

   (b)

   (c)

2 Study the *Caribbean: Population and Language* map on page 29 of your atlas.

(a) Follow the instructions to complete the map of Hispaniola provided.

   (i) Insert the City populations (2015) symbols.

   (ii) Use the *Haiti, Dominican Republic* map on page 46 of your atlas, to correctly label the names of the cities and towns that are represented by the City populations (2015) symbols.

(b) Are most cities on your map coastal or inland?
_____

(c) Give ONE piece of evidence to support your answer in **2(b)**.
_____
_____

3 **Study the map of *Dominica* on page 52 of your atlas.**

   (a) Describe the location of Roseau and the other important towns shown on the map.

   _____
   _____
   _____
   _____

   (b) Discuss TWO physical factors that have influenced the location of these towns on the island.

   _____
   _____
   _____
   _____
   _____

4 **Study the *Jamaica: Population density, 2011* map on page 41 to compare and account for differences in population density between the following parishes. Use the terms in brackets below to guide your explanations.**

   (a) St James and Trelawny (socio-economic)

   _____
   _____
   _____

   (b) Manchester and Portland (physical)

   _____
   _____
   _____

5. Reference the *Trinidad and Tobago: Population* map on page 66 of your atlas. Discuss how socio-economic factors have resulted in a high population density along the East–West Corridor of Trinidad.

_____
_____
_____
_____
_____
_____

6. State TWO characteristics of a primate city. Describe how the Caribbean's colonial history resulted in the creation of primate cities.

_____
_____
_____
_____
_____
_____
_____
_____

Population structure

1 Complete the following sentences. Use the words provided in the box at the end of each paragraph. Each word is to be used only once.

(a) A population pyramid provides information about the _____ of the population. The _____ are shown on the left side, while the females are on the _____ side.

right   gender   males

(b) The numbers on the left and right sides of the pyramids are age _____. Each group increases by _____ years. Therefore, at the _____ of the pyramid, ages zero to _____, are followed by ages five to nine directly _____. The groups continue _____ like this.

above   upwards   groups   four   five   bottom

(c) The numbers at the bottom of the pyramids are _____. These numbers must be read from the _____ of the pyramid at the bottom. You must start where the number zero is between the blue and _____ columns. Each full _____ represents 1% of the total population.

percentages   pink   middle   square

(d) Therefore, to calculate a percentage of _____, you start at the bottom middle zero and _____ the blue squares going left. To calculate a percentage of females, you start at the same zero, but this time count the _____ squares going right.

pink   males   count

2 Study the *population* pyramids on page 29 of your atlas to complete this exercise.

(a) Use the *population* pyramids to estimate the percentage in each category in the table below. Remember that one square represents 1%. Circle the correct answer from the set of choices provided.

COUNTRY	POPULATION	AGE GROUP	PERCENTAGE		
Haiti	Males ONLY	0–14	16%	8%	42%
Haiti	Females ONLY	0–14	4%	14%	16%
Haiti	Males and females	15–59	54%	26%	30%
St Lucia	Males ONLY	60–64	38%	2%	10%
St Lucia	Males ONLY	0–14	20%	12%	10%
St Lucia	Females ONLY	60–74	3%	5%	8%
Cayman Islands	Males and females	15–39	19%	32%	64%
Cayman Islands	Females ONLY	70–79	3%	22%	37%
Cayman Islands	Males and females	0–5	1%	5%	6%

(b) Which country has the highest percentage of school-aged children?

_____

(c) Which country has the most aged population?

_____

(d) What percentage of St Lucia's population is of working age (15–64) and what percentage are dependents?

(i) Working age:

_____

(ii) Dependents:

_____

(e) Which of the THREE countries would you expect to have the largest workforce in the future?

_____

(f) Give ONE reason to support your answer in part **2(e)**.

_____

_____

(g) Discuss TWO reasons why having a large working-age population can benefit a country's economy.

_____

_____

_____

_____

3 **Read the following paragraph.**

The workforce is the part of the population that creates income for a country. They support children and retired people directly by working as family units, and indirectly by paying taxes which help the government to keep the country running.

Decide whether each statement below is true or false. Add a tick to the correct column to indicate your choice.

STATEMENT	TRUE	FALSE
A population pyramid with a very wide top has a low number of retired people.		
A large workforce is good for a country.		
A population pyramid with a very wide middle and a narrow top and a narrow bottom has a large workforce.		
A population pyramid with a wide base has a large percentage of children.		
A high number of school-age children in a population right now is likely to result in a large workforce in the same population in the future.		
If a country's workforce is large but unemployment is high, citizens may turn to crime to meet their needs.		
A country with a large workforce and very few children will eventually have a large population of retired people and a small workforce.		

Population change

1 **Define the following key population terms:**

(a) Natural increase

_____

_____

(b) Infant mortality

_____

_____

(c) Birth rate

_____

_____

(d) Life expectancy

_____

_____

(e) Fertility rate

_____

_____

2 **Follow these instructions to draw TWO divided circles/pie charts representing the age groups of the male population in Jamaica, one for 1955 and one for 2015.**

Complete *both* tables on the next page to help you do this.

(a) Study the *Jamaica population* pyramids on page 41 of your atlas to select the correct percentage in each category. Circle your answer from the three options provided in the tables.

(b) Using your answer to part (a), calculate the total percentage of males in the population in the pyramid for each year.

(c) Calculate the angle you would use to represent each category in a pie chart.

(d) Use the circles provided to draw your pie charts with an appropriate title and key.

YEAR	GENDER	AGE GROUP	PERCENTAGE			ANGLE USED
1955	Males ONLY	0–14	19%	8%	2%	
1955	Males ONLY	15–59	29%	26%	24%	
1955	Males ONLY	60+	12%	2%	6%	
	TOTAL					360°

YEAR	GENDER	AGE GROUP	PERCENTAGE			ANGLE USED
2015	Males ONLY	0–14	20%	12%	42%	
2015	Males ONLY	15–59	11%	31%	80%	
2015	Males ONLY	60+	5%	3%	1%	
	TOTAL					360°

3 Complete the exercise below by circling the correct answer.

(a) Which statement below is true of STAGE 1 of the demographic transition model?

  (i) Death rates are high, birth rates are high.

  (ii) Death rates are high, birth rates are low.

  (iii) Death rates are low, birth rates are low.

  (iv) Death rates are low, birth rates are high.

(b) Dot maps are used to show:

  (i) population distribution

  (ii) population density

  (iii) population growth

  (iv) the demographic transition model.

(c) Which Caribbean country below has a low population density?

  (i) Barbados   (ii) Haiti

  (iii) Bermuda   (iv) Guyana

(d) Natural change in a population can be calculated by:

  (i) birth rate – death rate

  (ii) birth rate + death rate

  (iii) birth rate – immigration

  (iv) death rate – (emigration + birth rate).

(e) Which of the following is NOT a characteristic of a primate city?

  (i) Population at least twice that of the next largest city.

  (ii) A thriving economic core.

  (iii) It is usually the capital city.

  (iv) They are not usually found in Caribbean countries.

4 (a) For each stage of the demographic transition model, *circle* the word in the list provided which BEST describes the pattern of birth and death rates.

(b) Write ONE reason for the answer you selected at each stage.

(c) Name ONE country that is currently at each stage of the model.

Stage 1

Birth rate: high   low   increasing   decreasing

Reason: _____

Death rate: high   low   increasing   decreasing

Reason: _____

Country: _____

Stage 2

Birth rate: high   low   increasing   decreasing

Reason: _____

Death rate: high   low   increasing   decreasing

Reason: _____

Country: _____

Stage 3

Birth rate: high   low   increasing   decreasing

Reason: _____

Death rate: high   low   increasing   decreasing

Reason: _____

Country: _____

Stage 4

Birth rate: high   low   increasing   decreasing

Reason: _____

Death rate: high   low   increasing   decreasing

Reason: _____

Country: _____

Population growth: Caribbean

1. **(a)** Work from the *World: Country Statistics* tables on pages 174–181 of your atlas to fill in the missing data in the table below.

   Use the birth and death rate data to calculate the natural increase.

COUNTRY	BIRTH RATE (PER 1000)	DEATH RATE (PER 1000)	LIFE EXPECTANCY	NATURAL INCREASE (PER 1000)	POPULATION CHANGE (%)
Belize					
Haiti					
St Lucia					
Barbados					
St Vincent and the Grenadines					

   (b) Draw a labelled bar chart with an appropriate title to illustrate the data on natural increase you have calculated for the countries in part **(a)**.

   (c) Discuss ONE factor that may account for each of the following observations.

   (i) There is a similar birth rate between Belize and Haiti.

   _____
   _____

(ii) There is a similar life expectancy between St Lucia and St Vincent and the Grenadines.

(iii) There is a difference in death rate between Haiti and Barbados.

(d) In addition to natural increase, name ONE other factor that must be considered when calculating population change.

(e) Barbados has a lower natural increase than St Vincent and the Grenadines but a higher population change. Explain what this indicates about migration in the two countries.

(f) Suggest ONE advantage and ONE disadvantage of a high birth rate for a country's economy.

(g) Discuss ONE way in which access to education can affect the fertility rate of a country.

(h) Discuss TWO ways in which high unemployment can affect population change.

Population: China

1 Study the *China: Population growth rates* graph on page 128 of your atlas. Describe the general trends in each factor below.

(a) Total population

(b) Birth rate

(c) Death rate

(d) Natural increase

2 What was the Chinese 'one-child policy' and when was it introduced?

3 Explain TWO reasons why the 'one-child policy' was introduced.

4 Describe TWO incentives that were used to support the 'one-child policy' in China.

5 With reference to the *Population growth rates* graph on page 128 of your atlas, evaluate the effectiveness of the 'one-child policy'.

_____
_____
_____
_____

6 Explain ONE negative social effect and ONE negative economic effect of the 'one-child' policy on China.

_____
_____

7 Suggest ONE factor that may have led to the stabilisation of the death rate from 1975 onwards?

_____
_____

8 How would the stabilisation of the death rate affect life expectancy?

_____
_____

Population: Nigeria

1 State what the *Nigeria population* pyramid on page 112 of your atlas indicates about:

(a) the birth rate in Nigeria

_____

(b) the stage of demographic transition Nigeria is in.

_____

2 Study the *Nigeria Population growth* graph on page 112 of your atlas.

(a) Describe the general trend in population growth.

_____

(b) Describe the trend in the ratio of rural to urban population growth.

_____

3 Discuss ONE way a high rural population has affected the birth rate in Nigeria.

_____

_____

_____

4 State how health care standards have affected infant mortality and life expectancy in Nigeria.

_____

_____

_____

_____

5 Suggest ONE reason why dispensing health care is challenging in Nigeria.

_____

_____

_____

6 Nigeria has a history of ethnic conflict. Explain how this has influenced the effectiveness of government policy in regulating population growth.

_____
_____
_____
_____

7 Suggest ONE way in which population growth in Nigeria has been impacted by colonialism.

_____
_____
_____

Urbanisation

1 Define the following terms:

(a) urbanisation

_____

_____

(b) urban sprawl

_____

_____

2 Study the *World: Urbanisation* maps on pages 158 and 159 of your atlas to complete the exercise.

(a) Name the TWO largest cities in the United States of America.

_____

(b) Name the TWO largest cities in North Africa.

_____

(c) Name the largest city in South America.

_____

(d) Which continent has no cities with a population of over 5 million?

_____

(e) List the population densities of these cities:

    (i) Chicago _____

    (ii) Lima _____

    (iii) Lagos _____

    (iv) Mumbai _____

    (v) Shenzhen _____

(f) Use the *Largest urban agglomerations, 2015* bar chart to calculate the difference in population between:

(i) Tokyo and Osaka

_____

(ii) Karachi and Kolkata

_____

(iii) Manila and Jakarta

_____

(iv) Moscow and London

_____

3 **Explain THREE benefits of urbanisation.**

_____
_____
_____
_____
_____

4 **Explain THREE problems associated with urbanisation.**

_____
_____
_____
_____
_____

Urbanisation: Kingston

1 Study the *Kingston-Spanish Town conurbation* map on page 41 of your atlas.

(a) Explain how the map shows the influence of urban sprawl on the development of the Kingston-Spanish Town conurbation.

_____

_____

_____

_____

_____

_____

_____

_____

(b) Describe ONE physical factor that influenced the development of Kingston as an important urban centre.

_____

_____

(c) State the physical factor that has restricted the further development of residential areas to the north.

_____

(d) Explain ONE reason the Kingston/St Andrew commercial areas developed around the central business districts.

_____

_____

_____

(e) State TWO advantages of living in the Kingston/St Andrew residential area.

_____

_____

(f) What is ONE advantage of the location of Spanish Town's industrial areas?

_____
_____
_____

(g) Discuss THREE problems associated with urbanisation in Kingston.

_____
_____
_____
_____
_____
_____

(h) Discuss TWO methods that have been implemented to control urbanisation in Kingston.

_____
_____
_____
_____

Migration

1 **Define these terms:**

(a) Pull factor

(b) Push factor

(c) Positive net migration

(d) Negative net migration

(e) In-migration

(f) Out-migration

(g) Regional migration

(h) International migration

2 **Study the *Caribbean: History and Heritage* maps on pages 30 and 31 of your atlas to complete the exercise.**

(a) What was the reason for the forced migration of Africans into the Caribbean?

(b) List the TWO countries where most migrants to the Caribbean came from during the 1830s to the 1920s.

(c) Explain why people migrated from the Caribbean to Latin America.
_____

(d) Explain why people migrated from the Caribbean to Britain from 1945 to 1962.
_____

3 Study the *Intra-regional migration* map on page 31 of your atlas.

(a) List THREE countries from which the Cayman Islands receive a high number of migrants.
_____

(b) List TWO countries from which Trinidad and Tobago receive a high number of migrants.
_____

(c) List TWO countries that Guyanese migrants go to.
_____

(d) List ONE country that Antiguan migrants go to.
_____

4 Study the *Net migration rate, 2016* graph on page 31 of your atlas to state the net migration rate for the countries listed below.

(a) Cayman Islands
_____

(b) Antigua and Barbuda
_____

(c) Barbados
_____

(d) Guyana
_____

5 With reference to the *Cayman Islands Labour force and unemployment rate, 2010–14* graph on page 37 of your atlas, discuss why the Cayman Islands experience one of the highest rates of in-migration in the Caribbean.

_____

_____

_____

_____

_____

6 The agriculture industry in the Caribbean has been in decline due to external competition. Explain TWO ways this may influence intra-Caribbean migration flows.

_____

_____

_____

_____

_____

7 Access to tertiary education has increased across the Caribbean. Explain TWO ways this may influence international migration flows.

_____

_____

_____

_____

_____

Migration: Jamaica

1. Study the *Jamaica Net migration, 2002–2013* graph on page 41 of your atlas.

   (a) In what year was the lowest migration recorded?
   _____

   (b) In what year was the highest migration recorded?
   _____

   (c) Describe the trend in net migration in Jamaica.
   _____

2. Explain how the trend identified in question 1(c) can affect population growth.
   _____
   _____

3. (a) Which international country do most Jamaicans migrate to?
   _____

   (b) Explain how policy changes in the country identified in question 3(a) can affect Jamaican migrants?
   _____
   _____
   _____

4. Describe ONE route Jamaicans may use to migrate to North America.
   _____
   _____

5. State TWO reasons why Jamaicans may migrate to other countries.
   _____
   _____
   _____

6 Outline TWO of the challenges a Jamaican family may face when one or both parents migrate out of the country.

_____
_____
_____
_____

7 If migrant parents are successful overseas, explain what further impact this could have on the population structure in Jamaica?

_____
_____
_____

8 Describe what is meant by the term *brain-drain* and explain ONE negative effect it can have on Jamaica.

_____
_____
_____
_____

9 What are remittances and how can this positively impact the Jamaican economy?

_____
_____
_____
_____

10 Discuss TWO positive effects of migration on tourism in Jamaica since the 1990s.

_____
_____
_____
_____

Caribbean resources

1 Define the following terms:

(a) Renewable natural resource _____

_____

(b) Non-renewable natural resource _____

_____

2 To complete the map provided, study the maps on page 28 of your atlas and follow the instructions listed.

(a) Outline the exclusive economic zone of the following countries:

    (i) Barbados

    (ii) Antigua and Barbuda

    (iii) Cayman Islands

    (iv) Belize

- (b) Label the major container ports.
- (c) Label and add a symbol to the countries where oil refineries are located.
- (d) Label and add a symbol to the countries where bauxite mining takes place.
- (e) Label the country with the highest oil production.
- (f) Label and add a symbol to the THREE countries with the highest fish landings in CARICOM.
- (g) Label and add a symbol to the THREE countries with the highest roundwood production in 2014.
- (h) Add a suitable key to your map.

Economic sectors

1 Write definitions for each of the following terms and list TWO examples for each:

(a) Primary economic activity _____

_____

(b) Secondary economic activity _____

_____

(c) Tertiary economic activity _____

_____

(d) Quaternary economic activity _____

_____

(e) Economic profile _____

_____

2 Study the *World: Employment* map on page 169 of your atlas to compare the employment by economic sector of these countries:

(a) Madagascar and South Africa

_____

_____

_____

(b) Qatar and Egypt

_____

_____

_____

(c) China and Japan

_____

_____

_____

3 Explain why it is an advantage for a country to have a diverse economic profile.

4 Explain how manufacturing adds value to harvested natural resources. Provide an example to support your answer.

5 Discuss the relationship between colonialism and primary economic activity in the Caribbean.

6 Explain how international investment has been important for the processing and manufacturing industries in the Caribbean.

7 Explain how access to education has affected the Caribbean's economic profile.

8 Study the *Cayman Islands Economic activity* chart on page 37 of your atlas and the *St Kitts and Nevis Economic activity* chart on page 49 of your atlas.

(a) List ONE similarity and ONE difference between the economic profiles of the two countries.

_____

_____

(b) Suggest ONE reason for the difference listed in question **8(a)**.

_____

_____

9 Study the *Economic activity* charts provided to list ONE similarity and ONE difference between the economic profiles of the two countries. Suggest ONE reason for the similarity you have identified and ONE reason for the difference you have identified.

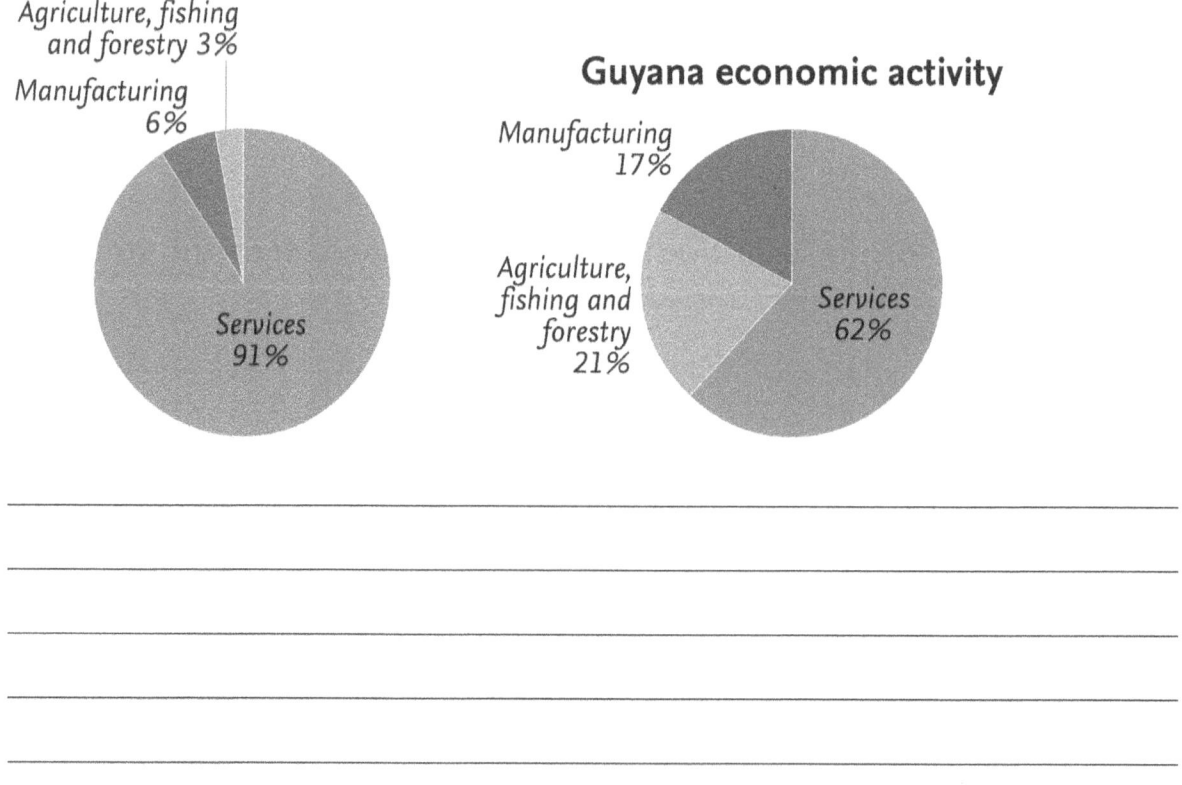

_____

_____

_____

_____

10 Suggest TWO reasons why few people are employed in the primary and secondary industries in the Caribbean while more people are usually employed in the tertiary industry.

_____
_____
_____
_____
_____

11 State how economic activity is important to each of these areas:

(a) Employment

_____
_____
_____

(b) Social services

_____
_____
_____

(c) Foreign exchange

_____
_____
_____

Jamaican resources: bauxite

1 Using the *Jamaica: Industry* map on page 42 of your atlas, complete the map provided below.

(a) Label the parish names (reference pages 38 and 39 of your atlas).

(b) Label the towns shown on the map on page 42.

(c) Shade the areas of bauxite mining and bauxite deposits.

(d) Label the major bauxite ports.

(e) Insert the railways.

(f) Insert the alumina plants.

(g) Add a suitable key to your map.

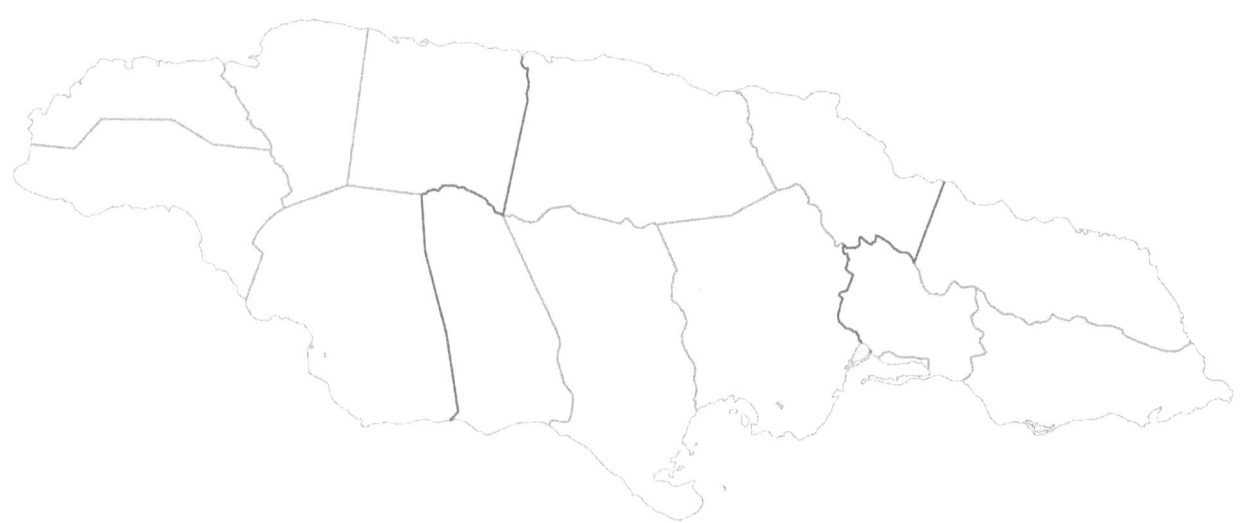

2 Name the FOUR parishes with the largest bauxite deposits.

_____

3 Suggest TWO reasons why not all the bauxite deposits are currently being mined.

_____

_____

_____

4 Explain the advantage of the railway system for the transport of bauxite and alumina.

_____

_____

5 Cross-reference the map on pages 38 and 39 of your atlas to list the TWO northern mountain ranges where bauxite deposits are located.

_____

6 Study the *Bauxite production, 1974–2014* graph on page 42 of your atlas to describe the trend in Jamaican bauxite production.

_____

_____

7 Explain how the following factors have influenced bauxite production in Jamaica:

(a) Closeness to North America

_____

_____

(b) International competition

_____

_____

(c) Government environmental policies

_____

_____

_____

8 Do some research about how bauxite mining and alumina production are connected and briefly state your findings.

_____
_____

9 List THREE uses of alumina.

_____
_____
_____

10 Discuss THREE ways that bauxite production benefits Jamaica's economy.

_____
_____
_____

11 Describe what is meant by the term *red mud*. State how it is problematic for the natural environment and humans.

_____
_____
_____
_____
_____

Trinidad resources: oil and gas

1 Using the *Trinidad: Mining and manufacturing* map on page 64 of your atlas, complete the map provided below.

   (a) Label the towns shown on the map.
   (b) Shade the factories zone.
   (c) Add symbols for the oil refinery, chemical manufacturing and asphalt mining areas.
   (d) Shade the Point Fortin and Gulf Oilfields and Hibiscus Gasfield.
   (e) Insert the oil pipeline connecting the Gulf Oilfields and Point Fortin to the oil refinery.
   (f) Insert the gas pipeline connecting the Hibiscus Gasfield to Point Fortin.
   (g) Add a suitable key to your map.

2 Cross-reference the *Trinidad and Tobago relief* map on page 61 of your atlas to suggest why the exploitation of other oil and gas resources in each of these areas is difficult.

(a) Northern Trinidad

_____

(b) Eastern Trinidad

_____

3 State ONE disadvantage and ONE advantage of using a pipeline system to transport oil and gas resources.

_____

_____

_____

_____

4 Explain ONE physical factor which led to the location of the petrochemical plant at Point Lisas and the oil refinery at Pointe-à-Pierre.

_____

_____

_____

5 Explain why exploitation of oil and gas resources on the west coast is easier than on the east coast of Trinidad.

_____

_____

_____

6 Study the *Trinidad and Tobago oil and gas production, 1970–2014* graph on page 64 of your atlas and describe the trend in oil and gas production.

_____

_____

_____

_____

7 Explain how the following factors have influenced oil and gas production in Trinidad:

(a) Closeness to North America

_____

_____

(b) Infrastructure

_____

_____

(c) Workforce

_____

_____

8 List THREE ways that gas is used in Trinidad.

_____

9 Discuss THREE ways that oil and gas production benefits Trinidad's economy.

_____

_____

_____

_____

_____

10 State TWO negative environmental impacts associated with oil and gas exploitation in Trinidad.

_____

_____

_____

Primary industry: agriculture

1 Reference the *Guyana: Georgetown* map on page 72 of your atlas to discuss THREE well-developed points about the importance of the canal system to sugarcane cultivation in Guyana.

_____
_____
_____
_____
_____
_____
_____
_____

2 Define the following key terms:

(a) Arable farming

_____

(b) Subsistence farming

_____

(c) Pastoral farming

_____

(d) Commercial farming

_____

3 State TWO ways in which the physical characteristics of the Caribbean have influenced the development of agriculture.

_____
_____
_____
_____

4. Describe TWO ways in which agriculture can harm the environment. For each, state ONE possible solution to this problem.

5. Explain THREE ways in which agriculture is important to Caribbean economies.

6. Compare sugarcane cultivation in Guyana and Brazil under the following headings:

   (a) Acreage

   (b) Farming practices

   (c) Labour

   (d) Technology

   (e) Markets

Primary industry: fishing in Belize

1 Explain why fishing is considered a primary economic activity.

2 With reference to the *Belize* maps on page 74 of your atlas, state THREE physical factors that may have contributed to the growth of Belize's fishing industry.

3 Study the *Belize: Farmed shrimp production* graph on page 74 of your atlas to describe the trend in shrimp production in Belize during 2003 to 2015.

4 List TWO species, other than shrimp, that are important to Belize's fishing industry.

5 What is the main export market for Belize's fishing industry?

6 Most of Belize's fishing community are part of a cooperative. Describe what a fishing cooperative is. State TWO advantages of working with a cooperative.

7 Describe THREE threats to fishing resources in Belize.

_____
_____
_____
_____
_____

8 Describe TWO ways in which fishing resources are being protected and conserved in Belize.

_____
_____
_____
_____

9 Suggest ONE way in which fishing in Belize is affected by international competition.

_____
_____
_____
_____

10 State ONE advantage the fisheries in Belize have compared to international suppliers.

_____
_____

11 Aquaculture has become an important aspect of Belize's fishing industry.

(a) Select TWO of Belize's natural features and explain how they are advantageous for setting up fish farming operations.

_____
_____
_____
_____

(b) State ONE barrier to setting up an aquaculture operation.

_____

(c) Discuss the advantages of aquaculture compared to traditional fishing in:

    (i) Quality control

_____

_____

_____

    (ii) Employment

_____

_____

_____

    (iii) Profits

_____

_____

_____

(d) State ONE way in which aquaculture can harm the natural environment.

_____

Primary industry: forestry in Guyana

1 Explain why forestry is considered a primary economic activity.

_____

2 Follow the instructions listed to complete the map of Guyana below.

(a) Reference the *Caribbean: Resources* map on page 28 of your atlas to shade the forested area.

(b) Reference the *Caribbean: Population* map on page 29 of your atlas to shade the area with the highest population density.

(c) Reference page 70 of your atlas to identify and label the Pakaraima Mountains.

(d) Reference the *Guyana: Features* map on page 71 to shade the Kaieteur National Park and label Kaieteur Falls.

(e) Label the Iwokrama Forest Reserve.

(f) Add a suitable key.

3. **Using page 71 of your atlas, answer the following questions.**

   (a) State the percentage of Guyana that was covered in forest in 2010.

   _____

   (b) State the percentage of Guyana that was covered in other wooded land in 2010.

   _____

   (c) State the percentage of wood that accounted for of Guyana's exports in 2015.

   _____

   (d) Identify which years during 2005 to 2015 have the highest and the lowest forestry production. State how many cubic metres were produced in the years you have identified.

   _____

   _____

4. **Describe TWO physical features of Guyana's forests that limit its use for logging operations.**

   _____

   _____

   _____

   _____

   _____

5. **List TWO types of wood produced by Guyana and ONE use of each type.**

   _____

   _____

6. **Describe how trees are harvested and transported in Guyana.**

   _____

   _____

   _____

   _____

7 Extended Learning

Do some research about how the Guyana Forest Commission manages forestry operations. Discuss TWO ways in which this is achieved.

_____
_____
_____
_____
_____

8 Extended Learning

Do some research about how the Iwokrama Forest Reserve helps with the conservation and sustainable management of Guyana's forests. Discuss THREE ways in which this is achieved.

_____
_____
_____
_____
_____

Secondary industry: food processing in Singapore and CARICOM

1 Use the data in the table on page 129 of your atlas to write a short paragraph comparing food processing in Singapore and Trinidad and Tobago.

_____
_____
_____
_____
_____
_____

2 Suggest ONE reason why Singapore has a very low level of primary economic activity.

_____

3 Singapore is an important global shipping and air travel location.

(a) Explain how its location has been an important factor leading to this development.

_____
_____
_____

(b) List THREE advantages this has for Singapore's manufacturing and processing industries.

_____
_____
_____

4 List THREE more economic activities that Singapore's economy is known for.

_____

5 Explain how Singapore's location has been an important factor for access to labour.

_____
_____
_____

6 State TWO of the issues faced by Singapore's workforce.

_____
_____

7 With reference to the food processing industry in Singapore, discuss:

(a) Sources of raw materials

_____
_____
_____

(b) Employee income and access to labour

_____
_____
_____

(c) Types of food products manufactured

_____
_____
_____

(d) TWO problems facing food processors

_____
_____
_____

(e) Government support for the industry

_____
_____
_____

8 Use named CARICOM examples and details to compare or contrast the food processing industry in Singapore with food processing in CARICOM, in the following areas:

(a) Access to labour

___

(b) Access to raw materials

___

(c) Size of individual operations

___

Tertiary industry: tourism

1 **Work from the *Jamaica: Tourism* graphs on page 43 of your atlas to answer the following questions.**

(a) Describe the trend in stop-over visitor arrivals from 2001 to 2014.

_____

(b) Identify the months in 2014 which had the highest and lowest stop-over visitor arrivals. Suggest why these figures may have been recorded in each of the months identified.

_____

_____

_____

(c) State which six months represent the peak cruise season in 2014.

_____

(d) State the main reason for stop-over visitor arrivals to Jamaica in 2014.

_____

2 **Work from the *Trinidad: Tourism* graphs on page 65 of your atlas to answer the following questions.**

(a) Describe the trend in cruise passenger arrivals from 2001 to 2014.

_____

_____

(b) Identify the month in 2014 which had the highest stop-over visitor arrivals. Suggest a reason for this occurrence.

_____

_____

(c) Calculate the total percentage of Trinidad's stop-over visitors (by country of origin) in 2014 that came from the Caribbean.

_____

3 State why tourism is considered a tertiary economic activity.

_____

4 Define a tourist.

_____

5 Draw a sketch map of your country and identify TWO locations/regions that are important to the tourism industry.

6 Explain why each factor listed below has allowed the Caribbean to develop a successful tourism industry.

(a) Location

_____

_____

_____

(b) Climate

_____

_____

_____

(c) Culture

(d) History

7 Discuss THREE ways in which a named CARICOM country has been successful in developing its tourism market.

8 Identify TWO challenges to the tourism industry in a named CARICOM country. State a possible solution for each challenge you have identified.

9 Compare and contrast the tourism industry in TWO named Caribbean countries.

_____
_____
_____
_____
_____
_____
_____
_____

10 Identify THREE ways in which tourism can have a negative impact on the physical environment of Caribbean countries. State ONE solution for each negative impact you have identified.

_____
_____
_____
_____
_____
_____
_____
_____

11 Describe ONE way that tourism can impact the social culture of Caribbean countries.

_____
_____
_____

Collins

THE SCHOOL BASED ASSESSMENT

THE SCHOOL BASED ASSESSMENT

What is the SBA?

The School Based Assessment (SBA) is a student project designed to test theories and concepts taught in an area in the real world. The project is based on a specific problem or issue. An SBA requires serious planning and organisation of two major components: a *fieldwork* exercise and a *report* on the findings of this exercise.

The fieldwork stage

Fieldwork is a critical component of the SBA and is multi-dimensional in its purpose. It is the deliberate interaction with our environment to gain first-hand facts (**primary data**), which is then guided by the work done already by others and published in textbooks, pamphlets and other articles (**secondary data**). A complete SBA involves three general categories or stages: *pre-fieldwork*, *fieldwork* and *post-fieldwork*.

Pre-fieldwork activities

1 **Decide on a topic:** This is perhaps the most crucial aspect of any SBA project, as it will form the basis of the entire study. Consider the following when choosing a topic:

- Ensure that it is relevant to the Geography syllabus. You can choose a topic from any of the specific objectives in the syllabus.

- Decide on a topic of interest and select a specific area of focus (which section of Geography interests you – *physical* or *human*?).

- If you are not sure about a specific topic, then try to identify a problem in your community or town to investigate. (*Remember to double-check its relevance to the Geography syllabus.*)

Example: Agriculture (Human systems)
Specific objectives of the Geography syllabus

> 17. locate areas in the Caribbean where commercial farming (both large-scale and small-scale) and subsistence farming are important;
>
> 18. compare the characteristics of large scale and small-scale commercial farming in a named Caribbean country;
>
> 21. explain the ways in which economic activities can contribute to environmental degradation in the Caribbean; and,
>
> 22. discuss measures to ensure the sustainable management of resources in the Caribbean.

2 **Compose an aim:** This gives the study direction and purpose and keeps the researcher on track. The aim of the field study should:

- be related to the topic of interest
- narrow the focus of the study
- be stated as a brief and simple question
- begin with a verb.

Example:

Aims of a field study
✓ To **identify** the location of the study area.
✓ To **observe** the farming methods used in the study area.
✓ To **describe** the characteristics of the farming methods observed.
✓ To **assess** the impact of hazards affecting farming activities.

Possible questions for investigation

1. What are the various farming methods used by farmers in the community of Flagaman, St Elizabeth?
2. What are the features of Dry Farming observed in the community of Flagaman, St Elizabeth?
3. What are the challenges faced by farmers in the community of Flagaman, St Elizabeth?

3 **Complete the strategy sheet:** This is a data form that you must complete, as it gives an overview of the study being undertaken by the researcher.

Example:

GEOGRAPHY FIELD STUDY STRATEGY SHEET

Candidate's name: Mary Sue   **Registration number** _____   **Class:** 10H

General topic of interest: Agriculture *[Choose a topic from the CSEC syllabus.]*

(a) **Possible question to be investigated:** *What are the features of Dry Farming in the community of Flagaman, St Elizabeth, Jamaica?*

(b) **Location of study area:** Flagaman, St Elizabeth, Jamaica. *[State the specific area (site) where the study will be done.]*

STRATEGY

A What is the aim of your study?
To describe the features of Dry Farming in the community of Flagaman, St Elizabeth, Jamaica.

B How will you obtain data? *[Give a general guide of how data will be collected for the study.]*
1. Use maps to locate the site.
2. Do field sketches and take photographs.
3. Observe the characteristics of the site.
4. Interview farmers, using prepared questionnaires.
5. Record farmers' comments (if permission is granted).

C Resources: *[State the specific instruments/equipment needed for the field study.]*
Questionnaires, maps, notepaper, clipboard, camera.

D How do you intend to present the data and findings in your report? *[Indicate the various ways the data collected will be presented in the report.]*
1. Sketch map showing the location of the study area.
2. Data will be represented with tables, graphs and photographs.

E Analyse and discuss data: *[Discuss the findings of the field study.]*
Outline and discuss the various features involved the process of Dry Farming.

Anticipated challenges: *[What are the likely challenges that you may face in the field?]*
The farmers may not be willing to be interviewed or recorded.

Possible solutions: *[How will you overcome these challenges if they should arise?]*
Talk to farmers to find ways to address their concerns about being interviewed.

Teacher's name _____   **Teacher's signature** _____

4 **Research:** Read information linked to the topic(s) you will focus on in the field. There are several geography-related sources of information available in both hard and soft copy that students can use to guide their research.

5. **Methodology:** The method used to collect information is important, as it provides critical details about the following:

- **How (the data was collected):** This describes the instruments used or that will be used to collect data for the SBA. These may include, but are not limited to:

    ✓ *Sampling:* This is a method of selecting a subset from a larger group to carry out research. It is not always practical to survey everything in a study area, because of its size; hence a sample is usually needed. Some of the more widely used sampling techniques include: *random*, *systematic* and *stratified*.

    ✓ *Surveys*

    ✓ *Photographs:* These can easily capture images that are sometimes challenging to explain with words and can also be used to help you sketch areas of the site studied.

    ✓ *Field sketches*

    ✓ *Recording of observations (patterns or anomalies):* This method involves the researcher actively engaging in the process of watching, touching, listening to and assessing human behaviour or the natural environment.

Example:

Record the observations made on the farm and farming activities

a. Size of plot (acres): _____

b. Land: Level ☐   Sloping ☐   Stony ☐

c. Crops: Present: No ☐ Yes ☐; Number of crops: _____

d. Describe the stage of crop(s): _____

e. No crops ☐   Land bare ☐   Land covered with Guinea grass ☐

f. Tools/machinery: _____

g. Irrigation: No ☐ Yes ☐; Drums ☐ Drip hoses ☐ Other: _____

h. Other observations: _____

✓ *Secondary resources:* This is data acquired through sources such as textbooks, websites, libraries, surveys or censuses and official government offices or divisions.

✓ *Questionnaires:* These can have **open-ended** or **closed-ended** questions. Closed-ended questions usually have options for the participant to select, for example a simple 'yes' or 'no', whereas open-ended questions require more thought to obtain further information. Lines to write the answer on are usually provided for these responses.

✓ *Interviews:* This is one of the most common methods of collecting data in fieldwork. Questions are asked to a person(s) – *interviewee(s)* – by the researcher (*the interviewer*) to collect data. Interviews can either be structured (*planned*) or random (*unplanned*).

Example:

INTERVIEW/QUESTIONNAIRE

1. How long have you been farming? _____ ← **Closed-ended question**
2. Do you own ☐ lease ☐ or rent ☐ the land you are farming on?
3. How many plots of land do you farm? _____
4. How much land do you farm? _____
5. What are the things you do when preparing to plant your crop(s)?
   _____
   _____

6. Why are you growing the crop(s) on your farm now?
   _____
   _____

7. What are the main problems you face, doing farming?
   (a) _____
   (b) _____

8. How do you solve the problems you face, doing farming?
   _____
   _____

9. If or when your crop is destroyed by natural or man-made occurrences, ← **Open-ended question**
   who provides help?

ORGANISATION	TYPE OF ASSISTANCE

10. Talking about the recent fire affecting this farming community:
    (a) How often is this area affected by wildfires? _____
    (b) What is the cause of wildfires? _____
        _____
    (c) What can you do to reduce the effects of wildfires on your farm? _____
        _____
        _____
    (d) What help do you think is most needed for people affected by the recent wildfires?
        _____
        _____
        _____

- **When:** The time the data is collected or will be collected is important and must be recorded. In addition, the length of time taken or will be taken to acquire the necessary data in the field should also be noted. This provides a timeline for data collected. Specific dates and time of day when the site was or will be visited must be mentioned.

Example:

> *The researcher(s) visited the farms in Flagaman, St Elizabeth on 30 September 2019, between the hours of 10:00 am and 4:00 pm. The interviews conducted with each farmer lasted approximately 30 minutes. A total of five farmers were interviewed.*

- **Where:** The study area, also referred to as the *site*, is significant to the field study, as it is the physical location where the data will be collected. Sketch maps are usually required; however, for the methodology simply state the specific area where the study was undertaken and the latitudinal and longitudinal coordinates, where possible.

Example:

> *The study was conducted in the farming community of Flagaman, St Elizabeth, Jamaica. The latitudinal and longitudinal coordinates for Flagaman are 17.8847°N, 77.7137°W.*

IMPORTANT NOTE

You are expected to conduct your fieldwork activity in an area or site that has been previously visited and properly assessed for its safety and suitability for research (usually by your teacher or supervisor). Failure to do so may put you as well as your classmates at risk of being physically hurt or may result in a weak study, owing to the lack of evidence. Never underestimate that environments can be quickly modified (*whether by human or natural causes*), without your knowledge.

Fieldwork activities

When collecting primary data, consider the following useful tips:

- Take all the necessary equipment and instruments that you will need to collect data. Examples of instruments include: *questionnaires, clipboards, pencils, camera, zip lock bags* and *loose paper*.
- When recording information, pay attention to what you observe or measure and record these in a clear and concise manner. For example, *the colour of the soil, the types of crops on a farm, the volume of traffic at specific times throughout the week*. If you are in a group, delegate portions of the work among yourselves to ensure that the data collected is accurate, comprehensive and sufficient. If you are working alone, you may have to multitask carefully and in an organised manner.
- Use your time in the field wisely. Some field studies are time-consuming; hence, you must make the most of the day spent researching. One way is to use data collection forms that do not require a lot of writing.
- Always follow the instructions of your teacher or guide. Do not idle or stray from your group or teammates.
- Always be polite, flexible and professional, especially in field studies where you have to interact with the wider population.
- Be respectful of the environment where you choose to carry out your research. It must always remain clean, intact and in the same condition as when you first visited.
- Be on time! If you are required to arrive at a particular location at a specific time, ensure you honour this timeline as people may be making a sacrifice to facilitate your study.

Post-fieldwork: preparing the field report

The **SBA report** consists of and follows the following format:

1 **Cover sheet:** Title of the report (*in the form of a question*), your name, registration number, school and year of examination.

Example:

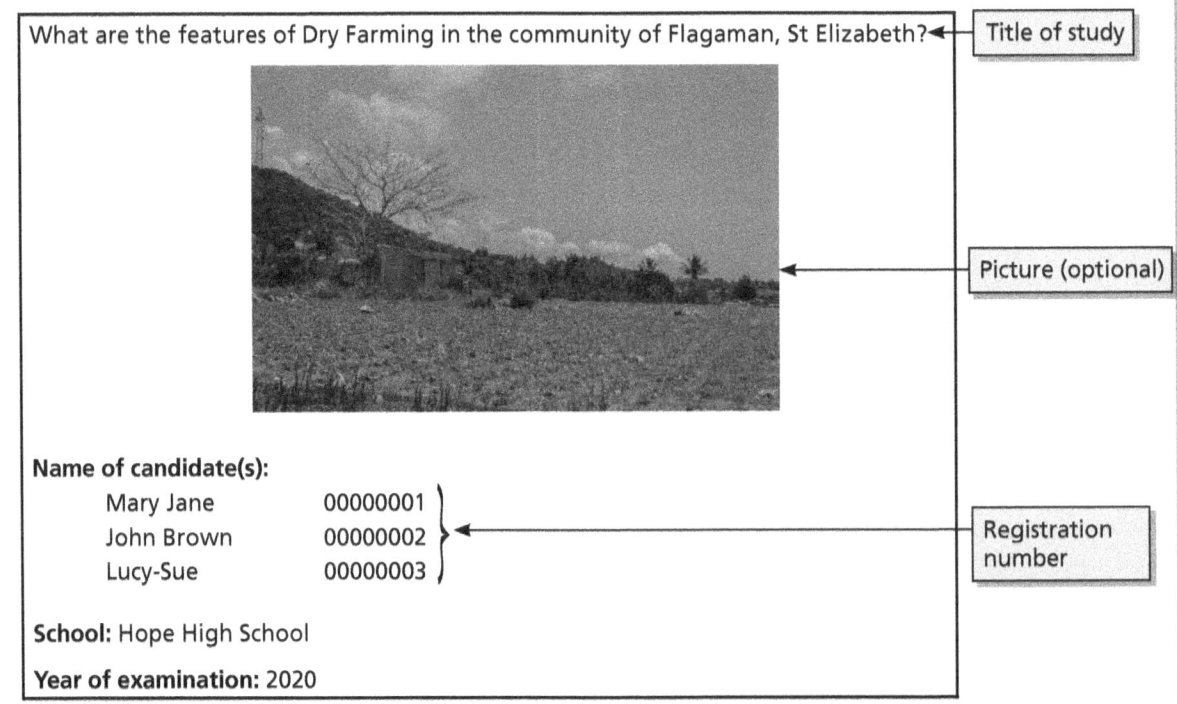

2 **Strategy sheet:** To be set on the second page.

3 **Table of contents:** Listing the sections of the report, with page numbers.

Example:

TABLE OF CONTENTS	
Contents	Page number
Introduction	1
Aim of the field study	2
Location of study area	3–4
Methodology	5
Data analysis and presentation	6–10
Conclusion	11
Bibliography	12
Appendix	13–15

4 **Introduction:** Background to the topic.

Example:

> According to The National Geographic Society, the term 'agriculture' is the art and science of cultivating soil, growing crops and raising livestock. It includes the preparation of plant and animal products for people to use and their distribution to markets. Our area of study is Flagaman, St Elizabeth, in Jamaica among a group of farmers. This location was selected because of the unique farming technique used by farmers, as well as the accessibility of the area to the team of researchers. This study therefore focuses on the characteristics of Dry Farming in this small community.

5 **Aim of the field study**

> What are the features of Dry Farming in the community of Flagaman, St Elizabeth?
>
> Or
>
> The aim of the study is to identify the features of Dry Farming in the community of Flagaman, St Elizabeth.

6 **Location of the field study:** Sketch maps and description of the study area(s). At least TWO sketch maps should be included:

✓ a *small-scale* map showing the area of study, with respect to the entire country

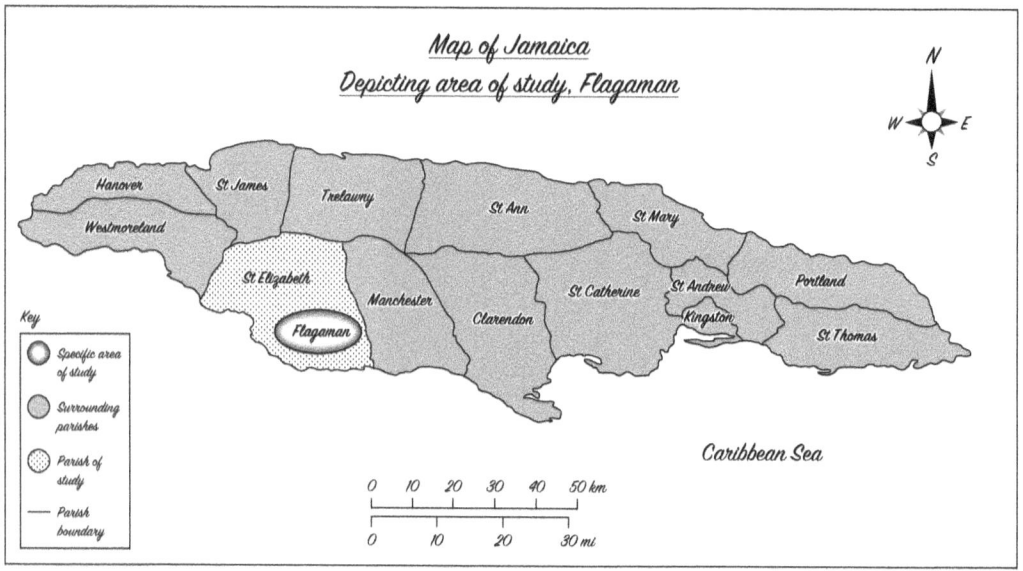

✓ a *large-scale* map, which shows the study area in relation to its immediate environs.

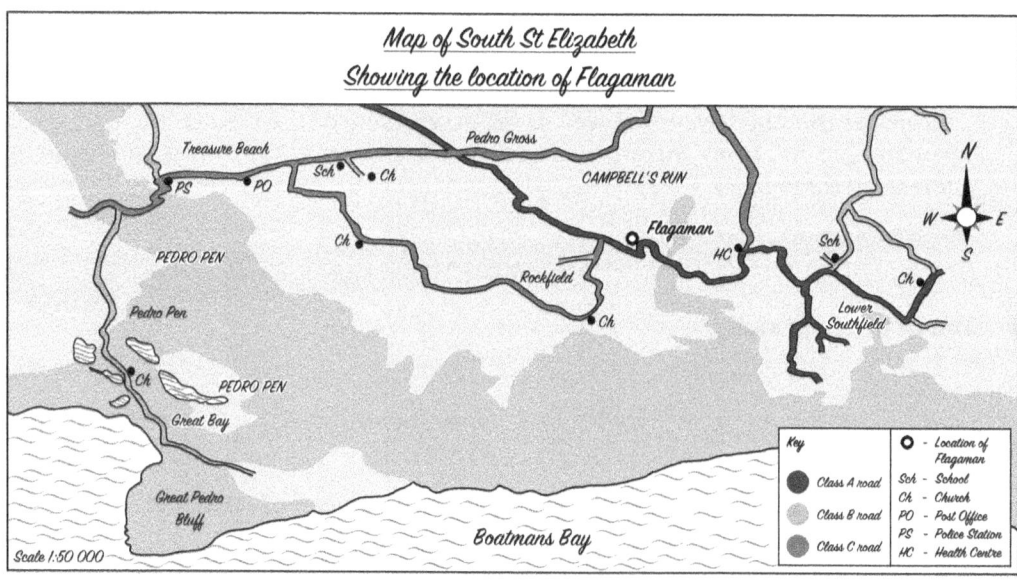

Location maps must always:

✓ include the five essential elements of a map: key, scale, title, border and north arrow

✓ be neatly hand-drawn and labelled appropriately. Avoid the use of electronically aided maps (whether computer-drawn or photocopied).

7 Methodology: Description of the data collection methods used; *how, where* and *when* the data was gathered.

8 Data presentation: Illustrations such as *maps, graphs, diagrams, field sketches* and *photographs* can be used to depict data collected in the field. At least THREE different types of illustration should be used when presenting data.

Example:

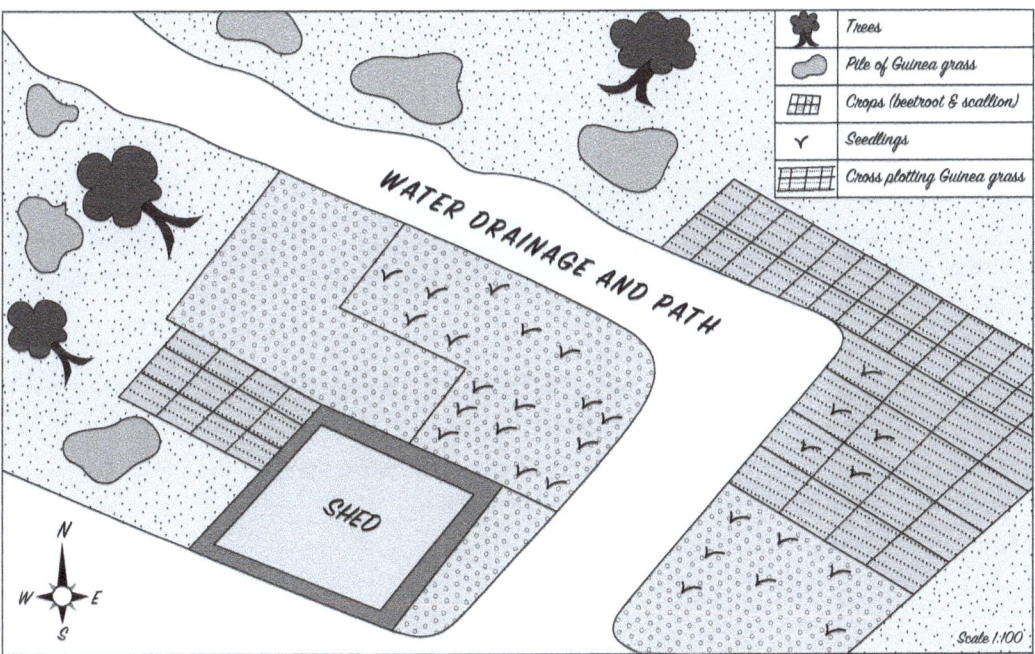

Field sketch of farm # 1 in Flagaman, St Elizabeth

Field sketches lend themselves to more detail and annotation of features observed in the field.

Tables are also widely used to organise data into categories that make viewing easy and the data accessible.

FARMERS INTERVIEWED	TYPES OF CROP GROWN
Farmer 1	Tomato and scallion
Farmer 2	Melon and cantaloupe
Farmer 3	Beetroot and scallion
Farmer 4	Thyme and scallion
Farmer 5	Tomato and thyme

Table 1. Crops grown by farmers in Flagaman, St Elizabeth

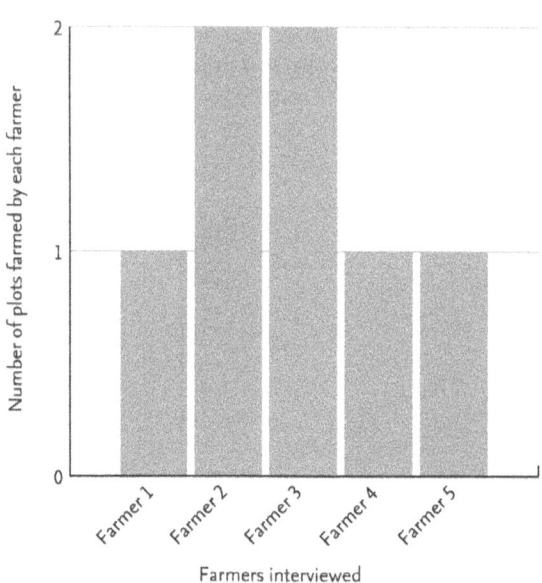

Figure 1

Graphs are commonly used to illustrate different types of data collected in the field. In addition to the bar graph in the illustration above, other types of graph include:

- ✓ pie chart
- ✓ histogram
- ✓ horizontal bar graph.

The type of graphs used are dependent on the type of study being undertaken. When using illustrations in the report, remember to include:

- essential elements such as border, titles and other labels, and keys, where necessary
- the source for graphs and maps acquired from secondary data sources
- the number for each graph or table that is used. Graphs, for example, would be **Figure 1**, **Figure 2**, and so on, and tables would be **Table 1**, **Table 2**, and so on.

If a photograph is used it:

✓ must be referred to as a '*plate*'

✓ should be properly annotated and cropped (*trimmed to remove any unnecessary features*).

Example:

Plate1. Showing how the Guinea grass is laid across the farm

Guinea grass overlying soil to retain the moisture.

Crops grown between the rows of Guinea grass overlying the farm.

The land is divided into squares, which are then cleared and then covered with the Guinea grass to retain as much of the moisture as possible. This is done throughout the growing season of the crops, until they are harvested. The grass is crucial for this special method of farming referred to as Dry Farming. Farmers sometimes venture outside the parish, to neighbouring parishes, to find this precious grass for their farming activities.

9 **Analysis:** This describes what data was determined from the fieldwork and how that data compares to the relevant theories or concepts. The analysis is usually presented in an integrated format with the illustrations, to provide a visual aid to the text. The analysis must also be related to the aim of the study and provide logical and consistent data.

Example:

Land ownership among farmers in Flagaman

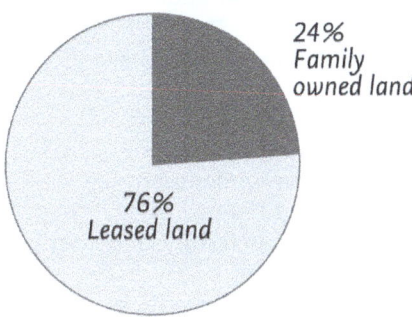

24% Family owned land

76% Leased land

Figure 2

All of the farmers interviewed grow an average of two crops per farm, either on family land, which represents 24% of the farmers, or land leased to them by the Anglican Church in the community, which accounts for 76% of the farmers interviewed.

10 Conclusion: A summary of findings, and whether or not the aim was achieved, to include recommendations and possible solution(s) to any problems identified in the study.

Example:

It can be concluded that Dry Farming involves several features, which are widely practised among farmers in the small community of Flagaman. One of the main reasons this unique type of farming is done is because of the peculiar location of the farming community, which is in southern St Elizabeth, a rain shadow zone. Consequently, farmers resort to this farming technique to maintain soil moisture and extend the growing season of the crops planted. Five farmers were interviewed, all of whom use the Dry Farming technique and are engaged in the process on a daily basis to ensure that their livelihoods as farmers are sustained.

11 Bibliography: A list of sources of information used to prepare, execute and make sense of the findings. These are usually listed in alphabetical order. There are several different styles of writing bibliographies, such as APA style, MLA, Chicago, and so on.

An example of an APA style is

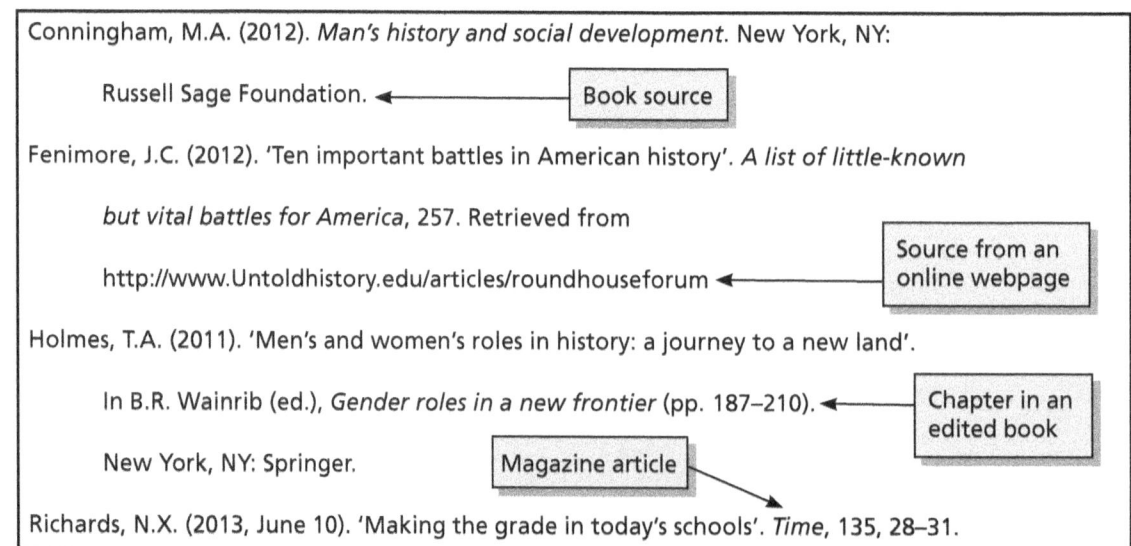

Conningham, M.A. (2012). *Man's history and social development*. New York, NY: Russell Sage Foundation. ← Book source

Fenimore, J.C. (2012). 'Ten important battles in American history'. *A list of little-known but vital battles for America*, 257. Retrieved from http://www.Untoldhistory.edu/articles/roundhouseforum ← Source from an online webpage

Holmes, T.A. (2011). 'Men's and women's roles in history: a journey to a new land'. In B.R. Wainrib (ed.), *Gender roles in a new frontier* (pp. 187–210). ← Chapter in an edited book
New York, NY: Springer.

Richards, N.X. (2013, June 10). 'Making the grade in today's schools'. *Time*, 135, 28–31. ← Magazine article

12 **Appendix:** If necessary, you could include: glossary, interviews, questionnaire and associated tally sheet.

Common errors made in SBA

Aim
- not related to the syllabus
- no study area identified
- vague

Location
- unedited digital maps
- incomplete and inaccurately drawn maps

Methodology
- inappropriate
- inadequate
- insufficient

Presentation and analysis of data
- not sequenced
- no supporting primary data
- no evidence of research

Conclusion
- rambling
- not connected to aim
- does not include main points from the finding(s)

Bibliography
- inadequate entries
- incorrect format used

The SBA is a mandatory part of the final score. Although it is worth 20 per cent of the final mark, if students do not complete and submit their report for the SBA then they will not receive a final mark for the exam, rather they will obtain a score of **U** or **UNGRADED**, as their exam scripts will not be marked. The activities on the following pages will help you practise some of the skills that you need to show in the SBA.

WORKSHEET

1 A map of Kingston (see appendix) should be used to answer questions 1(a) and 1(b).

Imagine you are planning to study fishing activities in the Greenwich Town Fishing Village, Kingston. With reference to the map, complete the following sketch maps in questions 1(a) and 1(b).

(a) **Figure 1** shows an outline of Jamaica and the parish boundaries. Using the map of Kingston provided, show the position of the Greenwich Town Fishing Village.

Figure 1. Location of the Greenwich Town Fishing Village

(1 mark)

(b) **Figure 2** shows part of the map of Kingston between eastings 68 and 78 and northings 45 and 53 on the same scale as the map. The 20m contour line, the coastline and the main water course (Hope River) are shown.

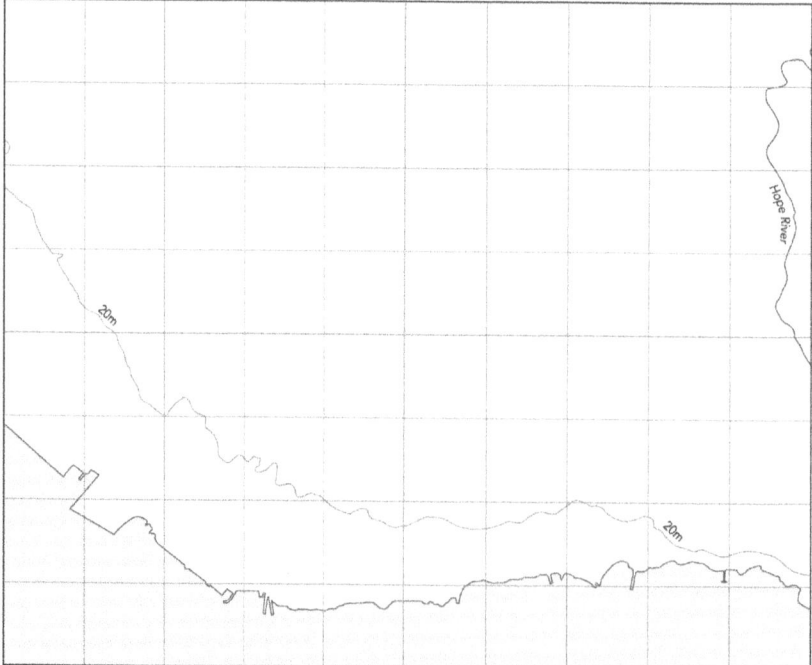

Figure 2. Site map of the Greenwich Town Fishing Village

Using the map in the appendix, add the following to the map on the previous page:

- ✓ the area named Jetty (*close to Greenwich Town*) **(1 mark)**
- ✓ the Class A roads and Dual Carriageways between eastings 45 and 49 **(1 mark)**
- ✓ the Kingston Harbour. **(1 mark)**

2. You are to study the extent to which fishermen in the Greenwich Town Fishing Village are challenged while undertaking their fishing activities.

    (a) Formulate a question or state a hypothesis to guide the collection of the data in the field. This will be the '*aim*' or '*purpose*' of your study.

    _____
    _____
    _____
    _____

    (2 marks)

    (b) As an introduction to your report, suggest a reason for studying the challenges facing fishermen in the Greenwich Town Fishing Village.

    _____
    _____
    _____
    _____
    _____
    _____
    _____
    _____
    _____
    _____

    (2 marks)

3. A number of instruments or equipment would be necessary to undertake your fieldwork on the day(s) you choose to visit the site.

    (a) List FIVE items for collecting information from the fishing village.

    _____
    _____
    _____
    _____
    _____

    (5 marks)

(b) Describe how and when you would conduct the research, which will form a part of your methodology.

How:

_____
_____
_____
_____
_____
_____
_____
_____

(3 marks)

When:

_____
_____
_____
_____

(2 marks)

4 During your field study you are likely to face challenges executing the field activities that you have planned.

(a) Identify ONE problem that you are likely to encounter in conducting the research (do not mention weather or illness).

_____
_____

(1 mark)

(b) State the approach you would take to resolve this problem.

_____
_____
_____
_____

(2 marks)

5 The data shows that the challenges faced by the fishermen have affected the size of their catch over a period of five years. **Table 1** shows the responses of the fishermen.

CHANGES IN SIZE OF FISHERMEN'S CATCH	RESPONSES (%)
Decrease	70
Significant decrease	10
Increase	5
No change	15

Table 1. Responses of fishermen about the size of their catch

(a) Using the data in **Table 1**, complete the pie chart below to show the responses of the fishermen in the Greenwich Town Fishing Village.

Figure 3. Pie chart showing changes in the size of fishermen's catch over five years

(3 marks)

(b) *'Pollution'* and *'environmental disturbance'* were identified as the TWO major challenges facing the fishermen in the Greenwich Town Fishing Village. Use **Table 1** to describe the effect these challenges could have on the size of their catch.

(6 marks)

6 You have recorded additional information on the amount of fish lost daily due to inadequate storage. The data is recorded in **Table 2** below.

NUMBERS OF FARMERS	AMOUNT OF FISH LOST DAILY (POUNDS)
6	10–30
4	30–50
2	Over 50

Table 2. The amount of fish lost daily due to poor storage

(a) What instrument would help you to collect the data recorded in **Table 2**?

_____

(1 mark)

(b) Use the grid below to construct a bar graph showing the data recorded in **Table 2**.

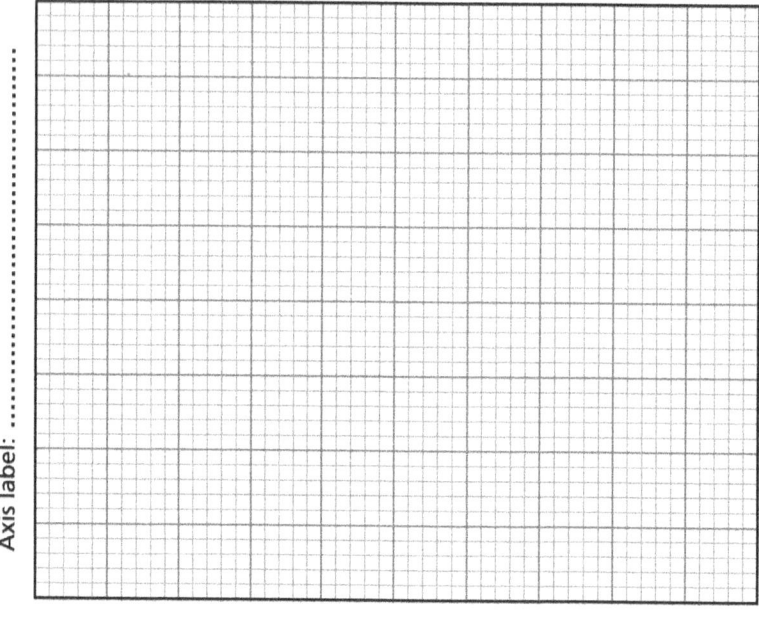

Axis label: ....

Figure 4. Amount of fish lost daily (pounds)

(4 marks)

(c) Explain TWO ways the loss of fish stock may affect farmers at the Greenwich Town Fishing Village?

_____
_____
_____
_____
_____
_____
_____
_____

(4 marks)

7 At the end of the report, a bibliography – list of references used – is required. List ONE of the elements of a book that is required for a reference entry in a bibliography.

_____

(1 mark)

APPENDIX

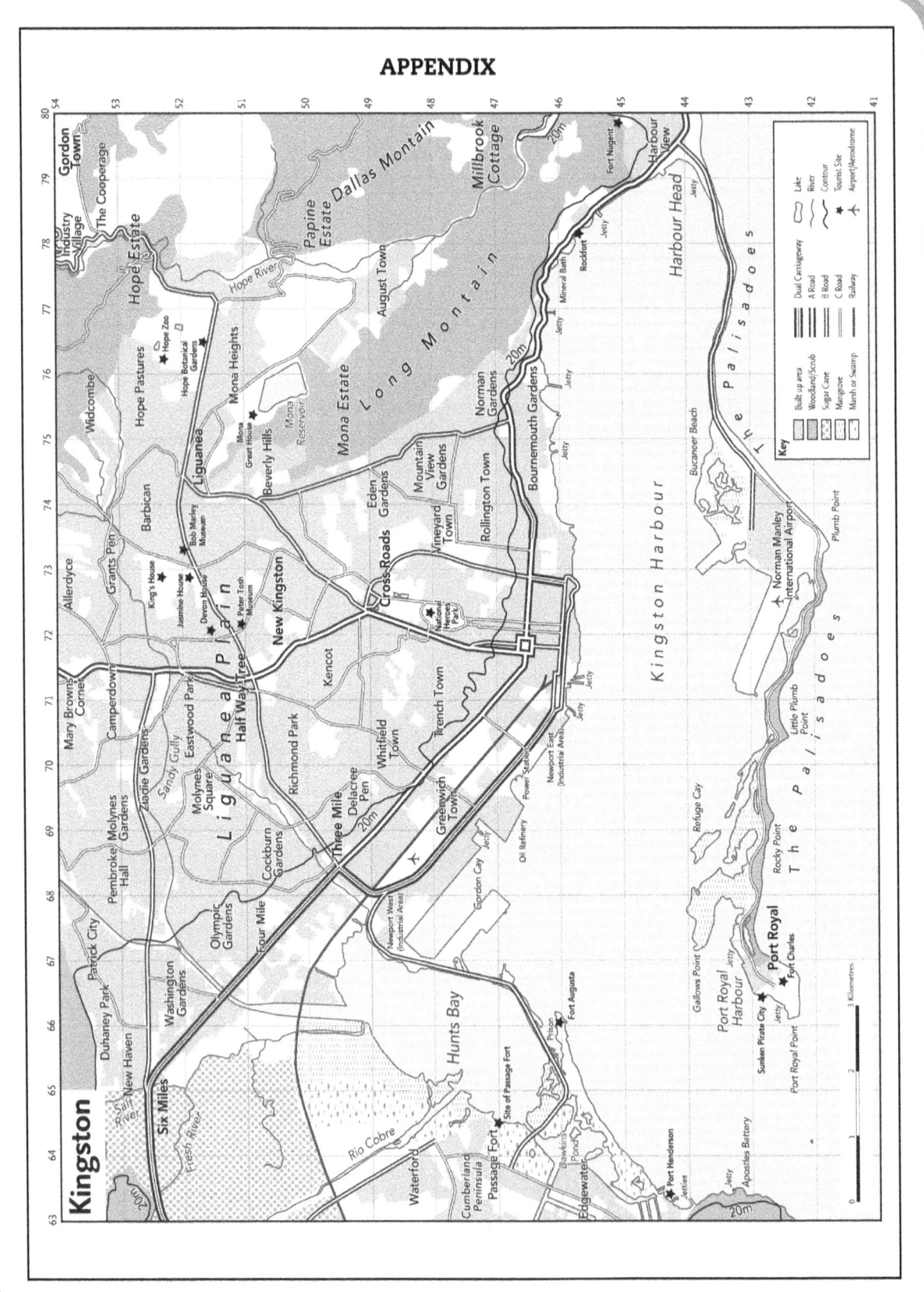

NOTES

NOTES

NOTES

NOTES